Body as Instrument

Body as Instrument

Performing with Gestural Systems in Live Electronic Music

Mary Mainsbridge

BLOOMSBURY ACADEMIC

NEW YORK • LONDON • OXFORD • NEW DELHI • SYDNEY

BLOOMSBURY ACADEMIC
Bloomsbury Publishing Inc
1385 Broadway, New York, NY 10018, USA
50 Bedford Square, London, WC1B 3DP, UK
29 Earlsfort Terrace, Dublin 2, Ireland

BLOOMSBURY, BLOOMSBURY ACADEMIC and the Diana logo are
trademarks of Bloomsbury Publishing Plc

First published in the United States of America 2022
This paperback edition published 2023

Cover design: Louise Dugdale
Cover image © Dean Lewins

A catalogue record for this book is available from the British Library.

Library of Congress Cataloging-in-Publication Data
Names: Mainsbridge, Mary, author.
Title: Body as instrument : performing with gestural systems in live
electronic music / Mary Mainsbridge.
Description: New York : Bloomsbury Academic, 2022. |
Includes bibliographical references and index. |
Summary: "Presents a range of design approaches for developing
motion-controlled digital musical instruments that reflect performer
perspectives and felt experience" – Provided by publisher.
Identifiers: LCCN 2021040176 (print) | LCCN 2021040177 (ebook) |
ISBN 9781501368547 (hardback) | ISBN 9781501369575 (paperback) |
ISBN 9781501368554 (epub) | ISBN 9781501368561 (pdf) |
ISBN 9781501368578 (ebook other)
Subjects: LCSH: Electronic musical instruments–Construction. |
Electronic musical instruments–Performance. | Gesture in music. |
Motion detectors–Design and construction. | MIDI controllers. |
Human-computer interaction.
Classification: LCC ML1092 .M24 2022 (print) | LCC ML1092 (ebook) |
DDC 786.7/1923–dc23
LC record available at https://lccn.loc.gov/2021040176
LC ebook record available at https://lccn.loc.gov/2021040177

ISBN: HB: 978-1-5013-6854-7
PB: 978-1-5013-6957-5
ePDF: 978-1-5013-6856-1
eBook: 978-1-5013-6855-4

Typeset by Newgen KnowledgeWorks Pvt. Ltd., Chennai, India

To find out more about our authors and books visit www.bloomsbury.com
and sign up for our newsletters.

To Daniel and Nicholas

Contents

Figures

Acknowledgements

This book is shaped and inspired by artists, Laetitia Sonami, Atau Tanaka, Pamela Z, Julie Wilson-Bokowiec, Lauren Sarah Hayes, Stuart Favilla, Joanne Cannon, Mark Coniglio and Garth Paine, who generously shared their time, vision and unique insights during online interviews.

I am grateful for the support and guidance of Senior Commissioning Editor, Leah Babb-Rosenfeld, of the Bloomsbury Music & Sound Studies series, Editorial Assistant, Rachel Moore, the Bloomsbury Academic production department, and the input of anonymous peer reviewers, whose detailed feedback assisted in clarifying the key concepts and underlying argument of the book.

Denis Crowdy as a beta reader offered encouragement throughout the writing and revision process and invaluable suggestions regarding the organisation and articulation of ideas. The book gained form during Creative Ecologies Lab writing groups led by Julie-Anne Long, and through discussions with mentor, Julian Knowles and colleagues at the Department of Media, Communications, Creative Arts, Language and Literature at Macquarie University (MQ). Ethical clearance for the research was granted by the MQ Faculty of Arts Ethics Subcommittee. Funding for audio and computer equipment was provided by an MQ Research Infrastructure grant.

I am deeply appreciative of photographers, Dean Lewins, Lori Eanes, and Goran Vejvoda who kindly contributed images, and J. D. Young, for his video art and graphics work with Deprogram, joining close musical collaborators, Robbie Mudrazija and Meeghan Oliver, in ensemble improvisations.

The digital media and arts projects emerging from Creativity & Cognition Studios provide a context for this practice-based research, as do the contributions of the New Interfaces for Musical Expression (NIME) and Movement and Computing (MOCO) communities.

I reserve eternal gratitude for my exceptional son, partner and parents.

Introduction

The body is widely considered the original human instrument, capable of producing sounds through voice and movement. Breath propels internal sound outwards during speech, singing and whistling, while physical actions transfer energy to generate sound through an external instrument or the body itself, in the case of body percussion. Motion-based sensing technology such as camera tracking in gaming and mobile applications has become increasingly widespread, expanding and redefining common forms of movement expression and awareness. This book explores how systems influenced by performer actions, from digital musical instruments (DMIs) and sound installations to dance–music interfaces (Wanderley 2001), shape the motivations and perceptions of musicians engaging with and designing them. Interviews were conducted with artists who use their bodies to control sound with a range of motion-sensing technologies. Their approaches to making, adapting and performing with gestural systems over a long-term basis are explored. Insights drawn from my own performance practice featuring a gestural system as my primary instrument frame this investigation.

Gestural systems are pervasive in everyday life, found in public touchscreens, mobile phones and art installations. Many individuals regularly perform tapping and swiping gestures when operating personal handheld devices. In musical applications, gestural systems offer the potential to interface with digital technology in a way that represents individual movement characteristics and physical nuances, conveying expressive information that is often lost in the regulated movements associated with traditional controllers such as digital piano keyboards for laptop performance. For live electronic musicians seeking richer modes of expression, gestural interfaces offer opportunities to incorporate greater physicality in their performances.

The performers and composers interviewed for this book reflect the diversity of performance and design approaches in the field. They are among the early implementers of motion-sensing technology in live electronic performance.

Composer and sound artist Laetitia Sonami is renowned for her work with wearable instrument, the lady's glove. She shares her experiences of developing and performing with a personalized invention over two decades before it became a seamless extension of her anatomy. Her current system, the Spring Spyre, is further removed from the body. Constructed from magnets, coils and recycled materials, it incorporates insights gathered from an extensive performance history with the previous instrument. Atau Tanaka explores the notion of the human body as musical instrument in live performances and compositions that translate physiological data capturing electrical activity in the skeletal muscles to musical processes. Vocal performers Pamela Z and Julie Wilson-Bokowiec develop works that merge operatic and extended vocal techniques with gesture. Improvisers with bespoke systems and instruments, Laura Sarah Hayes and the Bent Leather Band highlight the role of play and experimentation in refining tactile gestural systems and related repertoire. Composers and designers Mark Coniglio and Garth Paine reflect on how technical and physical processes meet in collaborative and interdisciplinary audiovisual pieces. The nine artists interviewed reflect the diversity of methods performers apply to detecting movement and establishing meaningful links between sound and movement in motion-controlled music. Embracing a range of sensing methods and performance approaches, they provide examples of how professional musicians active in the field design or customize gestural systems that highlight the body.

The book also describes personal explorations of movement-based performance over the past decade. Drawn to the field by a desire to bridge vocal, instrumental and electronic sources in performance, I leverage common performer gestures to process acoustic sound signals and control digital sound synthesis, aiming to engage more openly with audiences beyond the physical limitations of screen-based computer hardware. To this end, I have experimented with several customized systems that use touchless body movements to process and generate music, performing at the International Sound and Music Computing conference in Cyprus, International Space-time Concerto Competition and Museum of Contemporary Art and Vivid Festival of Light, Music and Ideas in Sydney, Australia. I started by augmenting voice and piano performance with digital signal processing and looping before developing a spatial system operated through improvised whole body movement. Digital audio hardware and software are accessed remotely using a camera to capture motion, reducing the impact on my usual playing style as a vocalist and keyboardist. Each performance informs design improvements, leading to the evolution and refinement of the customized performance system I currently use as my main instrument.

Unlike an acoustic musical instrument, an electronic instrument consists of a gestural interface (or gestural controller) that receives physical input from a performer, and a separate sound generation unit (Wanderley 2001). Gestural interfaces translate incoming movement information in a way that is understood by a computer, using a range of hardware devices, from tablets and mobile phones to wearable and remote motion sensors. These devices act as a bridge between sound and action, detecting human movement and converting it into numerical data in order to augment audio or visual performance, explore links between movement and sound, and expand the palette of sounds and control methods available to musicians (Miranda & Wanderley 2006).

Although once the domain of research labs equipped with sophisticated and expensive motion capture systems, motion-sensing technology has become widely accessible since the 1990s. Every mobile phone is equipped with accelerometers and gyroscopes for measuring movement. This has led to a rise in gestural multimodal systems and DMIs that capture and interpret continuous human movement or individual gestures. Gestural systems as a subset of DMIs can complement a musician's existing instrumental practice by augmenting acoustic or electronic sources. They are driven by patterns or phrases of physical movement that occur deliberately or spontaneously when controlling musical systems, conducting, or during vocal and instrumental solo or ensemble performance (Leman & Godoy 2010).

Because of the separation between the controller and the sound-generating device, gestural interfaces, unlike traditional acoustic instruments, impose no physical constraints to regulate the types of gestures that control sound (Mulder 2000). Daniel J. Levitin, Stephen McAdams and Robert L. Adams (2002) argue that this separation offers an opportunity to rethink controller design beyond integrated musical instrument constraints (Levitin, McAdams & Adams 2002). To explore this opportunity, designers must confront the challenge of designing mappings that make sense to the performer and audience, allowing for musically expressive control (Bencina 2005). Establishing coherent connections between gesture parameters and sound properties (gesture-to-sound mapping) (Bevilacqua, Müller & Schnell 2005) remains a persistent challenge for artists and designers aiming to translate the nuances of human movement into sonic processes. Mapping strategies determine how performer actions influence the controls of a sound-generating process such as a synthesis algorithm.

Divergent approaches to mapping within DMI and gestural interface design research results in a broad array of possibilities for musicians to consider when designing or customizing their own gestural instruments.

Live electronic musicians can employ gestural systems to devise their own movement vocabulary and imprint their unique body signature on a live performance work once freed from the micro-movements (Cascone 2002) associated with computer use and the operation of buttons, knobs and sliders on hardware sequencers, analogue synthesizers and effects units on stage. Laptop, instrumental and vocal performers alike, across numerous styles and art forms, use various synthesis methods, including granular and physical modelling synthesis, in gesture-operated sound generation. Digital signal processing, looping and sample triggering methods are also commonly controlled through artist movements.

Gestural systems can be operated either through tactile gestures using touchscreens, or with 'open air' gestures that are not traditionally associated with music-making, except in the case of conducting. In their purest form, gestural systems are invisible and untouchable. Players can be seen tracing shapes in the air, with no object to touch, tame or hold. Non-tactile gestural controllers rely on remote sensing technologies such as near-field capacitive measurement, infrared, ultrasound and video (Rovan & Hayward 2000). Instruments operated with spatial gestures lack the same degree of physical feedback as traditional instruments equipped with frets, a skin or strings that can be plucked, bowed or struck. The first electronic instrument, the theremin, is difficult to master as there is no felt resistance like frets or keys to touch, compared with traditional acoustic instruments that provide a tangible form of feedback to reinforce the performer's learning (Ihde 2013, 108).

When making sounds 'in the air', musicians need to be able to calibrate their gestures in space to maintain consistent control over pitch, rhythm and timbre. The processes for learning a touchless instrument include developing and refining motor skills and awareness similar to that of a trained dancer to be able to navigate through space with nuance and accuracy. Performers require a heightened ability to detect position in space, or proprioception, the sense of movement and position that encompasses 'tactility and gravitational orientation through vestibular sensory organs as well as kinaesthesia' (Sheets-Johnstone 2010, 218). Kinaesthesia is derived from the experience of movement or bodily awareness (Sheets-Johnstone 2010, 218). It offers an internal form of guidance directed by the felt sensation of one's own movement. Dance researchers Lynne

Anne Blom and L. Tarrin Chaplin (1988) describe kinaesthetic awareness as a primary perception and self-awareness of the body in motion. This skill delivers greater understanding of movement dynamics characterized by qualitative features such as smoothness, intensity, swiftness and expansiveness (Sheets-Johnstone 2013, 21). In the absence of an instrument or object to interact with, the kinaesthetic skill of controlling the weight, direction and magnitude of their body movements (Mandanici & Sapir 2012, 2) help performers regulate their speed and energy in real time (Acitores 2011, 219). Like professional dancers, musicians who play touchless systems need to remember and internalize the spatial positions of their performance and recall and re-enact these movements without any external physical guidance.

One way for musicians to access kinaesthetic awareness is to nurture new forms of musicianship by leveraging existing instrumental learning techniques. Musicians usually develop a feel for a particular instrument, aided by visual imagination and muscle memory. According to Clemens Wöllner and Jesper Hohagen (2017, 83), 'the most direct match between music and bodily movements lies in the kinaesthetic feeling in performance movements'. Pianists, for example, establish a sense of 'touch' through kinaesthetic sensations linked to different memorized tone colours, distinguishing them from other players (Doğantan-Dack 2011). These internal feelings are strengthened by kinaesthetic imagery during performance, or mental models of envisaged movement patterns and postural positionings, built up gradually through repeated performance experiences. This body memory and internal mind map of associations between gesture and sound direct a musician's movement, reflecting the integration between formalized training and intuitive interpretation, commonly referred to as 'feel'. While this internal form of guidance is vital for both instrumentalists and live electronic musicians, for players of touchless systems, these individualized maps become even more essential for ensuring precision and consistency in performance, fostering new ways of moving and inhabiting the body.

Defining and designing for the felt dimension

While there is much research into motion capture and gesture–sound mapping techniques, less is known about how performers engage with gestural interfaces and the types of skills needed to achieve nuanced and fine-grained control of existing or adapted systems. Current movement-based design research is primarily supported by empirical methods for studying movement through observation using motion capture technology. Much analysis of performer movement has so far focused on functional aspects – informing definitions of sound-producing and non-sound-producing gestures responsible for expressivity, excitation and modulation (Cadoz 1988). Gesture recognition allows a computer to interpret human gestures, converting them to commands. If a real-time gesture matches one the gesture recognition software has been trained to recognize, it executes the command linked to that particular motion. This method builds on functional perspectives of performer movement.

Additional attention to the underlying qualities and meanings of movement can provide valuable insights to inform gestural system design based on the motivations and innate body knowledge of the performer. To improve the precision and nuance of motion-based instruments, it is necessary to understand not only functionality but also a feeling-based approach to movement-controlled DMI design. Researchers in the broader interaction design realm, Calkin S. Montero et al. (2010) and Julie Rico, Andrew Crossan and Stephen Brewster (2011) have underlined the importance of acknowledging the feelings of gestural system users in design. Observing and analysing these feelings can assist in designing satisfying gestural systems and defining socially acceptable gestures in movement-based interaction (Rico, Crossan & Brewster 2011). Performing unfamiliar gestures in public settings, for example, can evoke feelings of discomfort, negatively affecting a user's experience. Understanding the feelings underpinning such interactions is therefore necessary for creating effective designs that promote satisfying and meaningful experiences.

The first part of the book reflects on the interplay between performance and design imperatives in gestural system development. It examines the current state of gestural musical instrument design, outlining common design criteria and prototyping strategies that involve the body and leverage existing physical understandings of performers. This overview of the design field provides a context for the decision-making process that musicians crafting and adapting systems undertake. Part One begins with an exploration of gestural manipulation of sound synthesis, vocal and instrumental augmentation, as well as interactive dance and art installations. The body as instrument metaphor frames this discussion about the functions and key design criteria of gestural systems within live electronic music. The role of body as a source of knowledge is then explored, widening prevailing functional and communicative interpretations of musical movement with first-person design perspectives emerging in the broader interactive arts field. The first part concludes with a discussion of the key phases involved in preparing gestural systems for performance.

The second part summarizes the diverse approaches that performers adopt when implementing and evaluating borrowed, adapted or self-designed systems and instruments in performance. In a series of interviews with long-term practitioners in the field, Part Two traces the performance and design methods these artists have conceptualized and refined, from selecting a movement-sensing method to deciding which particular movements to detect and how to translate these actions into sound. First-hand accounts of artists and designers throughout the book consider the unique movement vocabularies and techniques that evolve from this process. The performers and composers interviewed reveal the range of body data that can be captured, from full-body movements and hand gestures to electrical activity of the muscles. System types include augmented vocal and instrumental systems, wearable sensors and artist-authored software combining sound, video and movement control. The featured artists also explore a range of tactile, visual and purely auditory feedback modalities. The interviews focus on how performers develop the skills and techniques to deliver dynamic and convincing performances with gestural systems. Kinaesthetic awareness and developing a sense of feel in relation to the instrument emerge as essential skills for establishing mastery in the field of movement-based musical practice. A cyclical process of design, performance and reflection is also vital to continually advancing designs in line with personal movement and aesthetic preferences.

The final part on synergy and transformation examines the intersection between performance and design, exploring the effects of long-term engagement with gestural systems on performer agency and identity. The open-ended architecture of gestural systems can cater to different bodies and make musical performance more accessible to a range of body types, ages and abilities (Mulder 2000). In customizing their own instruments, musicians, like dancers, gain kinaesthetic skills, increasing awareness of how they move. In movement-based performance contexts, these novel and immersive movement experiences can stimulate new-found identities among contemporary musicians. Gestural systems provide the means to control processing like delay and reverberation, enabling vocalists to determine their overall sound and gain technical agency rather than relying solely on sound engineers, who do not share the same intimate connection with their sonic character and vocal capacity. Direct control over electronic augmentation, shaped by musicians' own hands and bodies, offers the opportunity to physically shape both the generation and processing of their individual sound – increasing agency in relation to live sound production and reinforcement. An expanded sense of agency can broaden the self-perception and identities of musicians as sonic artists and designers.

The implications of these altered performance practices are relevant to wider applications of gestural control, from health settings to virtual reality simulations and gaming. Musical applications provide insight into the challenges of understanding and analysing real-time performer movements, and the sensory fusion of visual, auditory and kinaesthetic perception that occurs when performing with gestural instruments, offering insight into how humans relate to their bodies and how their senses interact. These discoveries intersect with research in musical performance and the neighbouring disciplines of computer science, anthropology, philosophy and neuroscience. Insights emerging from live musical contexts can inform new knowledge in the related areas of human–machine relations and human movement potential and expressivity. The following part explores the philosophical foundations and cultural aspects of gestural system design for musical performance. It outlines the main phases involved in gestural system design, summarizing the many decisions musicians undertake when entering into a field largely characterized by customized approaches.

Part One

Invisible instruments

Gestural systems for musical performance

While there is a vast body of literature devoted to the design of gestural systems, less is known about how these systems are used in long-term performance practice. This section examines the decisions musicians initially face when joining the field, providing a context for the artist interviews presented in the second part. It explores definitions of gesture and outlines the stages involved in designing gestural systems for performance. Technical and creative practitioner perspectives on mapping human movement to sound are compared, revealing a multitude of approaches that potential designers and performers new to the practice must navigate.

Gestural interfaces cover a broad range of devices, from tablets and mobile phones to wearable and remote sensors. Gestural research in human–computer interaction (HCI) aims to widen the available gestural repertoire for users of computer systems. In conventional WIMP (windows, icons, menus, pointer) systems, the body's movement range is largely ignored. Motions are restricted to small-scale gestures that are highly repetitive, minimizing the physical inclinations and imagination of the user. This has led to an array of societal health problems associated with stationary technological activities including prolonged sedentary office work, passive web surfing and gaming (Kjölberg 2004, 353).

The increased dependency on laptops in live electronic music since the 1990s has also reduced the movement range available to musicians. The appropriation of a tool originally designed for office use in musical performance lacks the visual spectacle and theatrical codes that usually accompany musical performance, obscuring the cause-and-effect relationship between performer gestures and sonic outcomes (Cascone 2002, 4). The result is a displacement from the audience, Kim Cascone (2002) argues, in which the authenticity of the laptop performer/DJ is often questioned, rendering the transaction more of a broadcast than of a performance. The roots of this separation lie in the technologies and techniques characteristic of modern computer music, where the body is historically relegated to a secondary role in performance (Roddy & Furlong 2013, 3). Dating back to the advent of Musique Concrète, pioneered by Pierre Schaeffer in the 1940s, sounds have been removed from their source.

The device-centric nature of live electronic music, with its reliance on hardware sequencers, analogue synthesizers and effects units, has introduced a type of virtuosity based on manipulating an electronic signal with control knobs, switches and sliders (Ponce 2007, 47). The visual and corporeal elements of performing with computers and digital technology are largely obscured, leading to a re-evaluation of performance approaches that reflect the effort of the performer (Schloss 2003, 239). A shift to new paradigms of expressive interaction with machines in the music computing field followed, leading to a proliferation of systems that rely on gestures to generate and process musical signals and control hyperinstruments (Machover 2004) or virtual instruments. Within this context, immersive multimodal environments (MEs) facilitate multimodal user interaction that incorporates auditory, visual, tactile and kinaesthetic elements, operated using full-body movements, dancing and singing (Castellano et al. 2007). These interactive multimedia platforms provide tools for exploring the artistic potential of gesture and working with sonic, visual and haptic data (Leman and Camurri 2006).

The modern digital instrument has inner workings invisible to the audience, hidden within the circuits and virtual connections where 'physical energy is transduced via digital mappings' (Magnusson 2019, 7). The invisible part of an electronic instrument encompasses many parts and processes, including sensors, electronic circuits, software and an output system (Baalman 2017). Gesture and movement make these connections visible (Schloss 2003), as do screen projections of live coding processes and visualizations of movement and sonic data. The phases involved in creating these systems include selecting sensors to capture motion data and deciding what types of movement information to capture. Representing the richness and subtlety of human movement digitally is a complex task for both performers and designers.

Gesture offers a convenient way of analysing human movement by organizing it into smaller units (Mewburn 2009). Unlike physical movement, which can be objectively measured, gesture is a self-contained segment or unit of action (Jensenius et al. 2010, 19). Although discrete gestures do not accurately represent the continuous streams of motion of everyday movements (Mailman & Paraskeva 2013, 37), they offer manageable chunks of information that assist in identifying salient features, motifs and patterns of movement when designing

about their bodies – how to move, how to feel, how (finally) to *be*' (McClary 1998, 87).

Yet the problem of instrumentalizing the body rests in treating it as an observable presence (Márquez Segura et al. 2016), an approach pervasive in DMI design. In a creative context, Shusterman (2012, 1) argues that the body is often viewed from a functional perspective as 'a mere physical object for artistic representation or a mere instrument for artistic production'. Drawing on Shusterman's perspective, Jungmin Grace Han (2019, 41) questions the notion of body as instrument promulgated in Western music culture from another angle, suggesting an artificial separation that positions the mind as the 'leader' of the body. Influenced by the Cartesian mind/body divide introduced by philosopher René Descartes, classical music prioritizes technical motor skills, a view that 'overlooks the essential role of the body as a transformative subject, that is capable of creating and actualising an artistic ideal into living sound' (Han 2019, 42).

The theory of embodied cognition represents an alternative view, highlighting the primacy of bodily experiences in shaping thought. It represents a reaction to an earlier computational model of cognitive science, in which the mind served as the mechanism for converting experiences into abstract concepts divorced from the senses (De Souza 2017). Embodied cognition theory evolved from the works of philosophers Edmund Husserl, Martin Heidegger and Maurice Merleau-Ponty and cognitive scientists including Fransisco Varela, who oppose the separation of body and mind Descartes initiated. Embodied cognition theory arises from the phenomenological tradition founded by German philosopher Husserl, which values embodied experience over a disembodied notion of thought and knowledge. In his works, *Logical Investigations* (1901) and *Ideas: General Introduction to Pure Phenomenology* (1962), Husserl introduces phenomenology as a study of human experience. Departing from previous philosophical traditions that pursued universal and objective truths, phenomenology regards sensory and intuitive attributes as the core of consciousness (Nelson 2009). This emphasis on sensory and highly individual approaches makes phenomenology a popular framework for creative researchers in music (Nelson 2009, 58), particularly in the area of gestural musical instrument performance.

Husserl separates the body as a material object and the lived body, elevating the notion of a felt perspective. The lived body is ever present, reinforcing experience through felt sensations (De Souza 2017). Extending Husserl's concept of lived experience, Merleau-Ponty, in his influential work, *Phenomenology of Perception*

(1999), regards perception as the means by which embodied beings are situated in the world. The body explores the world through gestures, acting as 'our anchorage in a world' (Merleau-Ponty 1999, 144). It offers access to immediate experience prior to reflective or critical thought. Rather than being an object of the world, the body forms the basis of our communication with it (Merleau-Ponty 1999). Merleau-Ponty views the body as central to our understanding and involvement in the world: it assumes the basis for our experiences and belonging to an environment.

While previous musicological methods overlooked the body in music (Cusick 1994), embodied music cognition theory affirms that the body is central to our experience of music, underpinned by the notion 'that music is performed and perceived through gestures whose deployment can be directly felt and understood through the body, without the need for verbal descriptions' (Leman 2010, 127). The body's sensorimotor experiences, which refer to the integration of the sensory and motor systems, thus become the foundation of knowledge about musical performance and listening. Leman and Godøy (2010) emphasize the role of action in music by analysing the ways in which humans experience sound through their bodies. Experience is different for each individual, shaped by their unique perceptions and accumulated history of past experiences.

Gestural systems can establish new forms of connection between sound and body, reinforcing felt connections to bodily sensations beyond the physical constraints of a conventional instrument, which requires the body to conform to a particular design architecture. The absence of an external instrumental object can remove the distance of the body from sound creation, where once the object is plucked, hit or bowed, the sound is beyond the control of the musician (Magnusson 2019, 21). An increased focus on first-person perspectives of performers in the field has arisen from embodied philosophy, informed by a holistic perspective in which the body and mind are a unified entity. Musicians learn to think and adapt through the body, recognizing feelings that arise throughout performance. Moving consciously and with awareness is a skill not always emphasized in music education, compared with physically focused artforms like dance and theatre. Reflecting on her own transformative experience as a musician dedicated to inhabiting her body more deliberately, Han (2019, 50) views self-awareness as a pathway to self-transformation.

As we have seen, an instrumental view of the body limits our understanding of the transformative potential of movement-based performance. The next section examines the role of bodily experience in gestural system design and performance. How performers develop the skills necessary to effectively execute this unique type of performance with gestural systems is highly dependent on an awareness of the body's inner strengths, resistances and inclinations. Design methods that emphasize body consciousness are explored, reflecting the values and motivations of the artists engaging in them.

Body as source of knowledge

Our bodies, ourselves; bodies are maps of power and identity.

Haraway ([1985] 1991, 474)

Inscribed in movement is the individual body signature of the performer. American dancer and choreographer Martha Graham commented that 'movement never lies' (Graham, cited in Foster 1986, 28). This observation characterizes the search for natural ways of moving, a natural body, and natural choreography – ideas prevalent in American contemporary dance throughout the twentieth century, where dance became 'an outlet for intuitive and unconscious feelings inaccessible to verbal (intellectual) expression' (Foster 1986, xiv).

The notion of the body as a source of knowledge is prominent in contemporary dance research. Tuning into the inner feelings of the body is considered a core element of navigating dance practice in a creative and safe way. Similarly, participants in movement therapies are urged to let their body become a guide by recognizing and cultivating awareness of spontaneous bodily sensations.

Yet musicological research focused on the moving body in musical performance instead investigates the body as an objective presence that is analysed from a distant perspective. Empirical studies of Western art music performance concentrate primarily on the external appearance of the performer, revealing the contribution of instrumental gestures to conveying artistic intentions and visual cues about the structure of a piece to the audience (Davidson 2012). Motion-tracking studies of pianists' and clarinettists' performances demonstrate that ancillary or non-sound-producing gestures such as head and torso sway correlate with emotional expression, establishing a sense of timing, demonstrating structural transitions and promoting performer/audience communication (Wanderley 2002; Vines et al. 2004; Dahl & Friberg 2007).

These observational investigations echo movement study approaches in health and sports sciences, where motion capture systems identify movement anomalies in order to correct and improve elite athletes' movement performance. Personal fitness trackers that continuously monitor an individual's movement activities identify efficiencies and irregularities in both waking and sleeping states. Quantitative measurements of sleep patterns, resting and active heart rate, fat mass and daily activity levels are recorded through a range of wearable devices developed for general consumer use. Wearers of personal fitness gadgets are promised greater understanding and control over their health through access to these personal biometric statistics. Also popular in the security industry, biometric data in the form of unique fingerprints, voice patterns, facial and gait recognition can establish identity for authentication purposes.

Measuring biological data externalizes performers' experience of their own bodies. Movement is treated as a removed phenomenon, analysed in relation to criteria defined by designers. Movement is viewed as functional. A body's success is assessed in terms of fitting into an idealized body image and maintaining an optimum degree of efficiency. Specific diets and exercise routines involving targeted exercises performed at precisely timed intervals are recommended to tame the body to fit acceptable proportions of muscle and fat tissue. When it fails to meet these standards, more invasive surgical methods offer another option for reshaping and smoothing the body in accordance with societal conventions. This approach of precisely measuring and transforming the body reflects a broader cultural emphasis on external appearance, as evidenced by the performative act of selfie photography (Andreallo 2019). In a bid to conform to visual ideals, individuals are prone to physical strains and distortions from a lifetime of 'being assimilated to mental images of choice, necessity, value and inappropriateness', dance therapist, Mary Whitehouse (1995, 245) argues.

Alternative approaches to movement are evident in somatic disciplines such as Tai Chi, the Alexander Technique and yoga, as well as in movement therapy, which encourage movement that responds to inner impulses rather than a desire to appear acceptable or aesthetically pleasing (Whitehouse 1995, 250). Key proponents of somatic methods for self-discovery in the interactive arts, Richard Shusterman (2009) and media artist, Thecla Schiphorst (2009), assert that somatic practices can tune performers to their bodies through feeling. The field of somatics is primarily concerned with promoting movement awareness through an acknowledgement of felt sensations within the body, elevating first-person perspectives (Hanna 1988). Shusterman's pragmatist philosophy, somaesthetics, draws on the aesthetic notions of John Dewey ([1934] 2005) to enhance the body's efficacy and creativity through awareness of the feelings underlying movement, fostering 'the experience and use of the living body (or soma) as the site of sensory appreciation (aesthesis) and performative and creative self fashioning' (Shusterman 2020, 245). Shusterman also turns to somatic practices like Feldenkrais and Alexander Technique to pursue more satisfying experiences through enhanced somatic functioning. He aims to achieve an awareness of which movements are in harmony with the body and which movements hinder fluid, effective movement (Shusterman 2009).

Shusterman promotes somatic awareness as a method for improving action and performance by reconstructing habits and embracing different types of bodily behaviour. He characterizes 'soma' as 'the living, sensing, dynamic, perceptive body that lies at the heart of the project of somaesthetics' (Shusterman 2009, 133). In opposition to Merleau-Ponty, who favours spontaneity as an effective and uninterrupted way of allowing the body to guide us through everyday life, Shusterman argues that spontaneity must be balanced with conscious awareness in order to overcome any repetitive behaviour that may hinder experience (Shusterman 2009, 135). Whereas Merleau-Ponty describes spontaneous and habitual behaviour as primal, based on the body's capacity that 'guides us among things only on the condition that we stop analysing it and make use of it' (Merleau-Ponty 1964, 78), Shusterman (2009, 135) recognizes that, without critical reflection, these same habits can result in a damaging misuse of the body if movements are not executed in a manner that is harmonious with it. Acknowledging this somatic dimension is particularly critical in the context of gestural interface design, where the analysis of performer experience can inform a new understanding of relationships between movement and music.

The recognition that all human actions are embodied actions is central to recent trends in interaction design (Loke 2009) – sometimes considered the 'third wave' of HCI (Harrison, Tatar & Sengers 2007) – which incorporate aspects of ethnography, embodiment and phenomenology. Paul Dourish (2004) highlights the benefits of embodied interaction in leveraging innate human abilities to allow users to grasp an interaction without prior training. His comprehensive overview of embodied approaches in interaction design (Dourish 2004) is reflected in the research of movement practitioners in the field, including Larssen et al. (2007), Schiphorst (2009) and Loke (2009), drawing on ethnographically inspired and phenomenologically informed design, which emphasize the value of firsthand, first-person perspectives and experiential data in designing technology that incorporates human movement.

The value of exploring first-person experience is particularly relevant to designing and performing with gestural musical instruments, where feelings can act as a form of guidance. Many of the gestural systems explored in the second part of the book are directly informed by physical engagement and reflection on felt perceptions. This direct involvement in which performance and design intersect reveals information about how musicians feel movement, how they use movement as a form of expressive communication on stage and how they learn to consciously inhabit their bodies. In the next chapter, common design approaches within the field, which inform individual artist approaches, are outlined.

2

Design approaches

Many musicians who perform with gestural systems also design them, assuming the multiple roles of designer, researcher and user. They often leverage personal experience as a foundation for design. This chapter focuses on design approaches aimed at promoting discoverability as well as sensitive and subtle control of gestural systems during musical performance. Among these are techniques that capitalize on musicians' existing abilities, strengthening movement awareness and sensorimotor skills. This under-represented area of gesture and music research acknowledges the necessity to explore design approaches that encourage development of musicians' movement abilities in line with technical developments in gestural interface design. Strategies to support this aim are drawn from conceptual metaphor theory (Lakoff & Johnson 1980; Johnson 2007) and its applications in human–computer interaction (HCI), encompassing computer and cognitive science, embodied interaction design and interactive dance approaches.

By their hands-on nature, gestural systems encourage play and exploration (Saffer 2008, 19). Movement abilities and inclinations evolve through musical performance, triggering changes to identity, creative output, and self-perception. Musician and programmer, Thor Magnusson (2019, 12) characterizes this process as ergodynamics, where the performer develops skills in parallel with an instrument, resulting in a continuous dialogue between the two: 'This multi-dimensional discovery of an instrument is a dynamic process; it happens in time, and through it we find the object's power and potential (*potens, dynamis*) which, in turn, affects our ideas and embodied skills' (Magnusson 2019, 11). This complementary evolution is amplified in musical performance featuring gestural systems, where the musician's practice often influences design.

Developing movement-based digital musical instruments (DMIs) that reflect the unique body image and style of the performer enable individual movement vocabularies and styles to come to the fore. The highly personalized nature of these systems reflects low levels of standardization in the field. According to

Ryan Ingebritsen et al. (2020, 228): 'Just as language operates in different regions, cultures and individuals, interactive systems based on gestural controls are not easily translatable to other bodies and modes of performing and can lead to systems only designed for one performer, generally the creator.'

Despite the increasing affordability of sensors and the rise in commercial, academic and artistic gestural applications, lack of standardization in the field of gestural interface design means artists have few templates to follow and a limited number of commercial systems to purchase. A range of software and hardware toolkits are primarily available to support artists' design activities. Magnusson outlines the reliance of artists on pre-existing work in the DMI design area: 'In programming new instruments, we apply code libraries, compilers, synthesis and mapping algorithms, HCI guidelines, standards, and protocols such that we do not have to constantly reinvent the wheel' (Magnusson 2019, 223). Yet widely varying approaches to gestural system development result in divergent design priorities and criteria. The majority of gestural instruments are developed for specific projects and research contexts. The lack of cross-purpose applications may contribute in part to explaining why more musicians do not consider gestural control to be a viable alternative to acoustic instruments and DMIs modelled on conventional instruments.

In the general HCI field, Norman and Nielsen (2011) propose usability guidelines to meet the need for greater standardization of gestural interfaces. Norman (2010) believes more time is required to make gestural interfaces feasible alternatives to other interaction types, as the development of gesture as an interaction style is still in its infancy. His argument that further research is needed to decide how best to utilize gestures in interaction design and formalize conventions defining a standard set of gestures that retain the same meaning across a range of systems still holds true. Norman and Nielsen (2011) outline how abiding by fundamental principles of interaction design, independent of technology, can improve the usability of gestural interfaces. Although referring primarily to touch-based phone systems, their guidelines, which promote adequate feedback, discoverability and reliability, are also relevant to musical performance contexts.

However, general guidelines from the HCI field must be adapted in relation to DMIs (O'Modhrain 2011). Design criteria can vary according to the needs of a range of stakeholders, including performers, designers, composers, manufacturers and audiences (O'Modhrain 2011). Within the DMI field, there are several recurring design requirements. Evaluation guidelines proposed by Wanderley and Orio (2002) present learnability, explorability and controllability as essential prerequisites. Human factors researchers investigating HCI have assessed systems on the basis of 'naturalness', referring to the consistency and adaptability of the interface to the user's preferences. The incorporation of natural, uninhibited gestures ensures that people with different abilities, body types and skills can use them without overexertion. Within musical performance, this adaptability to individual body types and inclinations is often referred to as 'feel', or the fit of an instrument to a performer's body. Marshall and Wanderley (2011) evaluate the feel of an instrument according to the following characteristics: controllability, engagement, entertainment and potential for future performance.

Although it is an important criterion for general gestural interaction, in DMI research ease of use has been found to interfere with expressive potential because it does not push performance boundaries and challenge the performer to produce richer sonic results. David Wessel and Matthew Wright stress the need to balance ease of use during early adoption with strategies that stimulate ongoing interest and musical expressivity, a goal they describe as a 'low entry fee with no ceiling on virtuosity' (Wessel & Wright 2002, 12). Similar musical requirements guide the development of Steven Gelineck's (2012, 37) physical modelling instruments, including balancing simplicity of controls with infinite creative possibilities to achieve precision, expression and explorability.

A starting point for some designers is to adopt traditional acoustic instruments as a model or inspiration (Dobrian & Koppelman 2006; Kvifte & Jensenius 2006). Alexander Refsum Jensenius relates the laws governing acoustic instrument design to future instrument design (Jensenius 2013). Similarly, Tanaka (2000, 403) regards a successful sensor-based instrument as one that combines HCI design with acoustic digital lutherie. Turning to classical models of a performer's relationship with an acoustic instrument, Tanaka lists fluency, coherence and clarity as essential design characteristics.

However, acoustic instrument design differs from gestural interface design in one significant area – conventional instruments are usually designed to conform to average body measurements and skills, making them less adaptable to people with different dimensions and physical requirements:

> It follows that there is a need for musical instruments with gestural interfaces that can adapt by themselves, through "learning" capabilities, or be adapted by the performer, without specific technical expertise, to the gestures and movements of the performer. (Mulder 2000, 326)

Therefore, potential design criteria directed at improving performer engagement and satisfaction with gestural systems must also identify ways to make the instrument adaptable to varying performer needs, abilities and body types. The liberation from conventional control modes found in traditional instrumental architecture has manifested in a move towards greater inclusivity and diversity in DMI design. A growth in accessible digital musical instruments (ADMIs), which include adapted instruments and a range of control interfaces, such as brain–computer music interfaces (BCMIs), eye gaze and mouth-operated controllers, allow individuals with diverse abilities to participate in music making activities (Frid 2019).

Alessio Malizia and Andrea Belucci (2012) call for a high degree of personalization and customization in gestural interface design, enabling end-users to define their own gestural vocabularies to incorporate the cultural, human and contextual variations inherent in gestures. Gestural interfaces are often referred to as 'natural user interfaces' (NUIs), based on the assumption that they are controlled by everyday gestures and therefore require no prior learning (Malizia & Belucci 2012, 36). Yet many gestural vocabularies fail to represent users' true behaviour, as they are created in laboratory contexts (Malizia & Belucci 2012, 37). Malizia and Belucci (2012, 38) claim that current designs reflect a 'natural artificiality' where a set of gestures are imposed by a designer, when they should instead

> break down the technology-driven approach to interaction and provide users with gestures they are more used to, taking into account their habits, backgrounds, and cultural aspects.

Perhaps unconscious gestures, not involving hand movements, are more natural than the types usually used in interaction, they argue.

A common goal of gestural interface design is to provide users with the means to intuitively interact with an interface, so that no gestural vocabulary needs to be learned and gesture/action mappings are self-evident (Malizia & Belucci 2012, 37). The term 'intuitive interaction' generally relates to interfaces that a user can immediately utilize successfully (Antle, Corness & Droumeva 2009, 236) and that behave in a way that people expect (Spool 2005). Unlike a graphical user interface, or GUI, which enables every function to be discovered through the progressive exploration of menus, gestures cannot be easily represented in this type of visual format (Norman & Nielsen 2011). A gestural system relies on other techniques to promote intuitiveness. Similarly, Norman

(2010, 6) disputes the tendency to label gestural interaction as natural, arguing that 'gestures are neither natural nor easy to learn and remember. Few are innate or readily predisposed to rapid and easy learning.' The notion of natural gesture in music is also contentious, as any well-practised habit can become automatic (Sloboda 2005, 268). Through practice or rehearsal, a performer's own expressive repertoires can become intuitive or semi-automatic. In Alan Wexelblat's (1995, 180) interpretation, '*natural* means that the computer system adapts to the abilities and limitations of the human being, rather than the other way around' (emphasis added).

Levels of intuitiveness in a musical interface are also measured by how natural the control gestures feel (Overholt 2009). Yet many gestural systems often enforce a behavioural code that the performer must conform to, in the interests of discoverability. Dance-influenced interaction designers such as Jin Moen (2006) prioritize natural or intuitive movements when designing motion-controlled systems:

> In movement-based interaction we should provide possibilities for people to make use of their natural movements for communication and to create a dialogue with the system or application. When people can move freely and make use of their natural and spontaneous movement patterns, they can choose to use movements that feel good in the body and that correspond to the personal movement qualities. (Moen 2006, 14)

By focusing on spontaneous and improvised movement patterns, designers can equip users with an entry point into an interaction scenario, supporting an early sense of confidence and encouraging them to explore the interface further.

Moen's (2006) research has implications for the construction of intuitive connections between gestures and sounds through a gestural system's mapping strategy. Intuitive interaction has an impact not only on performer understanding of a gestural system but also on audience perceptions. Performer and choreographer Robert Wechsler, of the German-based dance company Palindrome, which developed the EyeCon video-based motion-sensing system, stresses the need to create intuitive connections between music, images and movement in performance: 'Straying even a little from what seems intuitive in terms of mapping – what *makes sense* on a feeling level – will result in a piece for which the outsider loses all perception of interactivity' (Wechsler 2006, 67). When artists directly engage in design, their gestural systems and instruments can be informed by their musical performance experiences and reflection on the felt dimension, leading to more nuanced works. Physically experimenting with a movement-based system allows performers to develop 'an understanding of its implications – the changes in the mindset and sensibility of artists as they put it to use' (Wechsler 2006, 75). Common musical activities like improvisation, rehearsal and performance offer ways to conceptualize and refine intuitive connections between movement and sound.

Experiential design

Interactive technologies controlled by body movement continue to grow – and, with them, design approaches that prioritize first-person, felt experience, which has expanded over the last decade since the third wave of HCI (Fogtmann, Fritsch & Kortbek 2008). Body-based design methods have risen in line with affordable motion sensors in mobile and gaming devices. How the body experiences and interacts with the world, rather than how it acts as an object to be seen and observed (Svanæs & Barkhuus 2020), has become a primary design focus.

Some artists conduct experiential prototyping, programming and refining original code while using a gestural interface. An exploratory approach contrasts with the more common technology-led practice of creating mapping strategies prior to implementing them physically. In the *Gesture≈sound experiments* (Bencina, Wilde & Langley 2008), a series of improvisations explore movement and sound interdependently, drawing out the dynamic relationships and complexities of the interconnected modalities (Wilde 2011, 34). Danielle Wilde, whose background in movement stems from physical theatre, notes that her fellow collaborators on the project, both electroacoustic composers, adopt a less physically oriented stance and ignore the expressive capacity of their movements. She observes that musicians and technologists appear to have less highly developed skills in expressive movement exploration (Wilde 2011, 35), perhaps due to a prevailing tendency to focus on the technical objects of performance systems rather than broader performance practice issues in existing design research (Green 2014). Yet a shift in focus from the technical functionality of a system to its creative potential is occurring (Torre & Andersen 2017).

Learning how to enact and describe the qualities of movement is considered essential to developing sensitive and expressive interactions (Erkut & Dahl 2017). As the above example demonstrates, musicians can collaborate with movement specialists from other disciplines and leverage existing musical experience to build on their prior movement knowledge. Some artists engage in movement improvisation, exploring new choreographies and gestural variations while developing or prototyping a system. From the dance and choreography areas, Loke and Schiphorst (2018) integrate somatics into design practice and theory, providing a practical template for exploring the potentials of the body in relation to interactive systems. They draw on early somatic principles espoused by Thomas Hanna (1998) and Eugene T. Gendlin, in his guide, *Focusing-Oriented Psychotherapy: A Manual of the Experiential Method* (1996) to formulate body focusing techniques for designers and artists. Gendlin introduced the term 'felt sense' to describe feeling that emerges from within the body.

Each artist has a different way of applying existing design ideas to their own practice. Artist-designer George Khut's collaborative design process starts with selecting the electronics and sensors, followed by a more active stage of wearing various sensors and testing a range of mappings linking bodily activities such as breath and heartbeat to sound and visuals, 'trying out different ways in which these mappings extend and transform my experience of these connections between body and mind' (Khut, cited in Candy 2020, 219). Khut aims to focus people's attention inward through his body-focused artworks. To achieve this, he first cultivates awareness of his own inner states, discovering how sound and visuals can amplify and transform personal body–mind connections.

Khut, Loke and Schiphorst turn to somaesthetics, focusing on attunement to the soma, or body–mind relationship, to explain the aesthetics of interaction (Shusterman 2013). Somaesthetic theory treats physical awareness as a means to self-improvement, drawing on body alignment practices like Feldenkrais and Eastern martial arts disciplines to understand the inner self. Interaction designer, Kristina Höök et al. (2016) incorporates soma-based design methods in collaborative brainstorming and design sessions where participants perform Feldenkrais exercises on a weekly basis, led by a trained Feldenkrais practitioner. This regular somatic practice triggered new ideas that informed design decisions for interactive works that encourage new forms of physical awareness, including the Soma Carpet and Breathing Light. The exercises stimulated a greater understanding of the types of bodily experiences participants could become immersed in. This direct experience with the designs also led to personal transformations: 'After a lesson, we all felt we had become more honest, more grounded in ourselves, more reflective, and a bit slower in our movements and reactions' (Höök et al. 2016, 3133). The interaction designers also realized the benefit of setting aside time for self-reflection to build a better understanding of the body, addressing unconscious habits which may or may not be beneficial (Höök et al. 2016, 3134).

Höök has recently turned her attention to the musical sphere, collaborating with researchers from the United Kingdom to prototype and workshop a breathing guitar (Avila et al. 2020). The adapted instrument is equipped with heat-sensing pads and inflatable pressure pads to promote awareness of movement and breath during guitar performance. Again, a Feldenkrais practitioner led exercises as part of an active and creative workshop-based ideation process to build designers' movement sensitivities. The collaborative project signifies a shift from a device-centric focus in the new interfaces for musical expression (NIME) community to foregrounding the body and embracing soma-based approaches in augmented instrument and gestural system design for live electronic performance (Mainsbridge 2021).

Shusterman (2009) recommends shifting the focus from one single body part to another to awaken an awareness of the relationships between different body parts and reflect on the uniqueness of each part. This perspective draws strongly on the Feldenkrais method, which traces the contributions of individual body parts to a unified whole. The somaesthetic appreciation design method (Höök et al. 2016) emerged from this philosophy. It is shaped by experiential techniques including exploratory workshops that incorporate Feldenkrais and experiment with a range of sensory stimuli including heat and light to find methods for guiding user attention throughout a planned somatic interaction. The underlying aim is to slow down the individual, encouraging introspection and the disruption of usual body habits. Somaesthetic appreciation design invites the user to dwell on the varying functions of related body parts and to linger in certain areas to avoid distraction and meandering thoughts (Höök et al. 2016, 3135). The Soma Carpet, which embodies these principles, is designed to steer focus to different areas of the body, much like a progressive relaxation meditation that involves tensing and relaxing individual body parts in turn.

When assessing physical experience, it is often done in relation to external physical phenomena or during social interactions. Shusterman (2009) asserts that no self-reflection can occur without this external context. Walking on varying surfaces, for example, can alter awareness of the body. Treading on solid concrete compared to stepping on loose grains of sand that compress and cushion the feet affects an individual's usual balance and gait, altering their posture and relationship to gravity and the ground. For musicians performing with gestural systems, lighting, sound reinforcement, room acoustics and audience participation also shape body-based understandings in live settings. A performer's practice can play a substantial role in shaping design, allowing the development of design imperatives that flow from the performer's inner desires and movement inclinations.

Paying attention to and experimenting with different combinations of movement parameters through movement improvisation also leads to increased sensitivity to felt sensations and an enhanced ability to produce and direct movement with greater subtlety and range (Loke 2009). Wilde (2011) integrates this theme into the design of *hipDisk*, a wearable interactive device that deliberately induces experimental and awkward behaviour to encourage users' playfulness and hone their movement awareness. In a similar vein, Schiphorst (2011) proposes an approach called attentional redirection to unearth insights from first-person experience as part of design workshops for the Whisper project, which explores new movement vocabularies emerging from interaction with wearable body systems. This technique draws on contemporary dance and meditation methods to redirect focus away from outer phenomena to inner felt sensations. In the next section I explore how this focus on the felt sense can influence common gestural system design decisions among artists.

Merging with the instrument

Designs that capitalize on musicians' existing physical patterns and skills, or body schema, may enhance exploration and satisfaction with gestural systems. A key tenet of embodiment theory espoused by Merleau-Ponty is that human motor responses are built up through motor memory that reflects natural habits and culturally acquired gestural routines (Merleau-Ponty 1999). These gestural routines, or body schema, describe an individual's intuitive understanding of their own body in relation to space:

> The theory of the body schema is, implicitly, a theory of perception. We have relearned to feel our body; we have found underneath the objective and detached knowledge of the body that other knowledge which we have of it in virtue of its always being with us and of the fact that we are our body. (Merleau-Ponty 1999, 206)

Preceding this argument, Heiddeger in *Being and Time* ([1927] 2010, 69) wrote about tool use, asserting that tools are ready to hand when they fade from awareness, eventually becoming an extension of the self. A common goal among many musicians is to reach a state where an instrument becomes a seamless extension of the body. When performer and instrument merge, the instrument is no longer perceived as a distinct entity but is felt from within, similar to how a vocalist experiences the internal vibrations of their vocal cords (Nijs 2017). The instrument becomes integrated into the somatic knowledge of the musician (Behnke 1995). Similarly, effective gestural systems are usually those perceived as transparent and integrated into a musician's performance practice.

A performer's body schema is influenced not only by unconscious habits but also by conscious awareness. Merleau-Ponty distinguishes between body image and body schema, describing 'the difference between a perception (or conscious monitoring) and the actual accomplishment of movement, respectively' (Gallagher 2005, 24). In contrast to body schema, body image denotes an awareness of the body in relation to the environment. The two concepts overlap when body image exercises an influence on the performance of body schemas (Gallagher 2005, 24–5). As a dancer practises extensively to gain proficiency, guided by conscious awareness of each motion (Gallagher 2005, 35), they reach a level where the movement is integrated into their body schema and can be performed without conscious reflection. Body image steers individuals' understanding and deliberate control of their bodies so that they do not simply perform habitually as automatons without any forethought. According to De Souza (2017, 17): 'The body image involves conscious awareness of my body, the lived body as intentional object. The body schema, on the other hand, is preconscious and supports automatic movements.'

Once a pianist has sufficient experience, they can recall entrenched physical patterns as their hands traverse the keyboard (De Souza 2017). Unlike novices, who must constantly look at their hands to correct their positioning, expert pianists have acquired kinaesthetic associations between particular body postures, fingerings and tone colours. When musicians reach this point, the body cannot be separated from the conventional instrument or gestural system, but rather 'the instrumentalist's musical capabilities emerge through interactions of body, world, technique and tool' (De Souza 2017, 17). Julie Wilson-Bokowiec, a trained dancer and skilled vocalist, has attained a high level of expertize with her main gestural system, the Bodycoder, allowing her to repeat pre-rehearsed patterns at will, even without adequate auditory feedback in certain live performance situations. She has learned to master her body movements and vocalizations in the absence of reliable speaker and monitoring setups.

Gestural musical instruments demand highly developed kinaesthetic awareness and proprioception, requiring musicians to acquire body control similar to that of a dancer (Pedrosa & MacLean 2008, 22), yet kinaesthetic training is rarely featured explicitly in traditional instrumental education. The onus is on musicians to independently learn how to refine their sense of proprioception to control instruments in a nuanced way. Merleau-Ponty defines two types of proprioception – conscious and preconscious (Tanaka and Donnarumma 2019, 81). Tanaka and Donnarumma ponder whether conscious proprioception can be absorbed through repetition by the performer, becoming preconscious, automatic and integrated into their body schema, even without formal movement training. This is where the performer assumes responsibility for consciously altering their body image, allowing themselves to broaden their physical abilities and potential for movement variation and regulation.

The way that the body adapts to an instrument contributes to experiencing flow (Csikszentmihalyi 1996). When the body itself becomes an instrument, its abilities and architecture are reflected in the performer's body schema, forming a tight somatic fit that places the performer into a flow state. The feeling of peak flow experiences evolve through long-term engagement with the instrument, shaping the performer's physical skills and even body shape. Richard Moore (1998) introduced the concept of 'control intimacy' to describe this match between the skills of an experienced performer and the desirable musical outcomes available from an instrument. This connection is most evident in vocal performance, as well as the violin, sitar and flute, where 'micro-gestural movements of the performer's body are translated into sound in ways that allow the performer to evoke a wide range of affective quality in musical sound' (Moore 1998, 22). Wessel and Wright (2002, 2), suggest that control intimacy can be achieved with appropriate control metaphors and low latency systems, encouraging users of new instruments to further develop their pre-existing skills and personal style.

Musicians receive sonic and kinaesthetic information during performance that they must respond to swiftly and also anticipate. Leman (2016, 134) characterizes musicians' need to develop sensorimotor schemes to effectively execute rapid and complex technical passages expressively as sensorimotor prediction. Using a process called chunking, musicians can assemble their performance actions according to groups of notes and recognizable patterns, making it possible to 'carry out a mental activity and bodily activity simultaneously' (Leman 2016, 135). While playing one passage, they can simultaneously prepare for the next. This frees the performer to focus on the imaginative aspects of their performance rather than only on the technical parts, according to Leman. He uses the example of scales, a rudimentary part of instrumental training, to illustrate his argument. As musicians recall these patterns during performance, a level of automatization takes place that lessens the cognitive load on the musician.

Designers who are able to leverage these familiar gestural schemes and habits through appropriate mapping and feedback can support movement exploration and the development of new spatial movement languages and sensitivities. Musicians can respond to kinaesthetic, auditory or visual representation when planning an action. Performers may alternatively refer to sonic images or kinaesthetic patterns to initiate and structure their actions (Leman 2016, 137). Leman notes that of all these forms of feedback, kinaesthetic data is 'generally rapidly accessible and reliable because it is based on neurones that sense the muscle activity' (Leman 2016, 137). Neurones that detect auditory, visual and tactile information are less reliable. However, combining more than one channel of information or multisensory perception can further promote a performer's sensorimotor learning (Leman 2016, 126).

Gestural systems that strengthen kinaesthetic feedback to improve a musician's physical and sensorimotor skills during performance give form to the instrument in the musician's imagination and developing body schema. Connections between physical gestures, the kinaesthetic sensations associated with them, and resulting timbral and melodic shapes can highlight potential connections between performer actions and resulting sounds in gestural instrument design, leveraging the tendency of musicians to use inbuilt mental representations and motor programs to guide their movements. Performers internalize motor imagery, establishing cognitive maps to structure their physical activities and act as a foundation for performing more skilled and challenging motor exercises such as executing a complex passage in virtuosic instrumental performance. This combines with a sense of 'feel', or the capacity of an instrument to respond to their actions, as well as the consistency of the relationship between actions they perform and the corresponding sounds that are produced (O'Modhrain 2000, 81–2). Feel can be built up from a combination of muscle memory and visual imagery over many years of training and practice (O'Modhrain 2000). Therefore 'feel' can be described as the fit of an instrument to a performer's body: 'The "how it feels" consideration is part of a feedback loop between action and instrument response' (Paine 2015, 84). The feel of a gestural system or instrument influences controllability, engagement, discoverability and ease-of use.

Kinaesthetic aspects are also highlighted in user experience considerations within embedded, tangible and wearable interactive product design (Svanæs and Barkhuus 2020). How a system 'feels' is at the forefront of design methods such as iterative design, prototyping and user studies, which aim to generate empathy with the user. The complexity of designing for the body involves balancing the connections between the user, product and context. When looking at live performance, it is clearly different to everyday use situations. The performer must feel comfortable moving and relating to the technology in new ways that may not always be immediately understood by the audience.

The interplay between performance and design is pervasive in the field of gestural performance. The next part explores how the performance activities of nine artists feed into their gestural system designs, movement awareness and body image. Each performer brings unique sets of performance approaches, combining a range of sensor and feedback types. These unique perspectives are influenced by diverse performance backgrounds and levels of movement expertize, resulting in varying techniques for performing with novel gestural systems as solo and ensemble instruments.

Part Two

Performer approaches

Instrument, technique, and sound coevolve, as elements of each are preserved or adjusted.

(De Souza 2017, 105)

To better realize the creative potential of gestural systems, it is important to reflect on performer approaches to gain a deeper understanding of how musicians develop a relationship with a gestural instrument or system over time. The second part of this book explores professional performer approaches to designing and working with gestural systems, investigating how an instrument can become integrated into a musician's overall practice and body schema. This investigation delves into the types of techniques musicians who use gestural musical systems develop to express themselves physically.

Interviews with nine long-term practitioners in the field provide insight into how individual artists interact with gestural systems on a prolonged basis and how this extensive experience has impacted their motor skills, identity and sense of agency. The participants represent a range of relevant career areas, including vocal and instrumental performance, production, software development and instrument design. The following chapters highlight practitioner experiences across three sensory and feedback modalities: kinaesthetic, tactile and visual. Some artists featured exploit haptic modalities, using sensors with vibrational force feedback, while others pursue vocal and instrumental augmentation through a combination of tactile, wearable and remote sensing systems. Whether artists choose to use haptic interfaces that offer resistance and a sense of tactility or wearable devices that act as an extension of the body is a deeply personal artistic decision. Atau Tanaka and Pamela Z, for example, extract intimate bodily information through electromyographic (EMG) sensors, which detect the minute movements of muscle cells, measuring the electrical activity of muscles as they move using electrodes placed on the skin. The EMG signal is an electrical voltage generated in response to the nerve's stimulation of the muscle, capturing inner muscular tension and effort, which influence the types of gesture recognition, sound synthesis mapping and performance techniques used.

It is impossible to approach body-based performance without engaging in the practice itself, Shusterman advises, advocating artists and designers embark on a practical enaction of somaesthetic philosophy (Lee, Lim & Shusterman 2014). My own experience in the area focuses on unencumbered gestural systems in which whole body movements are detected through computer vision. Engaging in this type of performance has awakened my interest in somatic practices like yoga, pilates and free-form dance. In the spirit of phenomenologist and dancer Susan Kozel (2007), my reflections on art and technology stem from first-person observation of feelings and impressions captured after live electronic performances with motion capture systems across a range of projects, pondering questions about the digitally augmented body like 'where my body ends and the synthetic body begins' (Kozel 2007, 214). In the final section of Part Two, I outline my own experiences throughout several performance projects with customized gestural systems over the past decade, culminating in a live album recording. I move beyond the technical issues of instrument design towards an exploration of embodied musical performance and improvisational practices combining visual imagination, evolving bodily awareness, and nuanced movement expertize.

As discussed in Part One, musicians who enter this field must firstly decide how and what type of movements to capture, before selecting methods for linking the data with specific sonic processes to establish relationships between movement and sound. Musicians also have the option to purchase specialized products like the Mi.MU smart gloves, designed in collaboration between pop musician Imogen Heap and University of the West of England in Bristol (Mitchell, Madgwick & Heap 2012) and inspired by Elena Jessop Nattinger's (2014) glove for vocal augmentation. The majority of performers in the field, however, develop customized systems, gradually acquiring the movement, musical and technical knowledge to make gestural systems suitable for their own purposes.

Like trained dancers, musicians using gestural systems require a degree of physical mastery 'to deliver a truly embodied performance with electronic music' (Schacher 2012, 199). However, musicians often lack formal movement training and must therefore attain these skills by leveraging movement techniques acquired during instrumental education or through direct engagement with the gestural interface, processing visual and proprioceptive feedback in order to orient and calibrate their performance gestures with sufficient precision. Instrumentalists, conventionally trained through repetitive exercises to develop techniques that become 'automatic, habitual routines that allow the musician to shift his or her attention to abstract musical ideas' (Tarvainen & Järviö 2019, 6), are typically less adept at spatial movement than dancers, for whom movement improvisation is a regular part of their practice. This elevates the necessity to hone movement abilities in parallel with conceptualizing and refining a gestural system.

Performers with little or no programming experience can find setting up software to control motion-sensing devices a complex undertaking, constituting a significant barrier to accessibility (Murray-Browne & Plumbley 2014, 213). To create a gestural performance system that reflects their personal movement style and preferences, performers need to be able to capture and interpret movement in a meaningful manner, coupled with a detailed understanding of their own movement potential. Apart from movement awareness and skills, performers who design or customize their own gestural systems also require a general understanding of music theory and programming, as Magnusson (2019, 59) suggests:

> In order to make the instrument, we need to know precisely what we require from the programming language, relevant protocols, the DSP theory, the synthesis theory, generative algorithms, and music theory, and we need to have a good understanding of human–computer interaction.

Collaboration is one common way of combining skill sets in different areas of movement and programming expertize. Mark Coniglio and Dawn Stopiello, founding members of dance and multimedia company, Troika Ranch; Suart Favilla and Joanne Cannon of the Bent Leather Band; and Julie Wilson-Bokowiec and Mark Bokowiec, co-designers of the Bodycoder system, compose and develop systems for collaborative live performances. Preparation in the studio involves physically experimenting with prototypes, refining movement vocabularies, assembling compositional elements, and calibrating responsiveness and dialogue with the system.

The artists in the next two chapters, Laetitia Sonami and Atau Tanaka, have an extensive history of designing and performing with self-designed instruments, cultivating unique physical identities that border on cyborg narratives rejecting distinctions between machines and the organic data of human motion. Sonami is renowned for her performances with the lady's glove, and more recently, the Spring Spyre, an instrument crafted from recycled objects. Like Tanaka, her latest works explore machine learning (ML) to develop sound-movement mappings and perform high-level analysis of gesture, assisting in the interpretation of sensor parameters to avoid complex programming, as even the simplest gestures can generate a substantial amount of movement parameters and complicated data parameters (Bevilacqua, Schnell & Fdili Alaoui 2011). Sonami provides examples of gestures and sounds such as the values of synthesis parameters to train a learning algorithm that creates a model based on linking specific performer motions and sound parameters (Fiebrink & Sonami 2020). Tanaka's biosensing system interprets energy steering the muscles, adding a layer of ML to evoke a dialogue with the machine.

Systems that can make future decisions based on predominant musical and motion patterns can challenge usual notions of control. This forms the basis of a discussion about the different possibilities associated with gestural system operation using implicit sound-movement mapping that relies on ML and explicit mapping approaches configured by the performer designer. After many years of performance, these idiosyncratic systems and instruments can assume a life of their own, introducing elements of unpredictability that reshape the performer's physical identity and creative choices.

3

Laetitia Sonami

For sound artist, performer and composer Laetitia Sonami, novel gestural instruments have the potential to transform their creators. Best known for her performances with the unique instrument, lady's glove (1991–2016), Sonami moved from wearable to object-based interaction for her most recent system, the Spring Spyre, constructed from a recycled steel ring and magnets. Such customized instruments 'change how we listen' and 'shift engagement', Sonami reflects, stressing the value and risk inherent in artist-led instrument design in terms of broadening creative potential and mirroring the unique physical signature of the performer.

Sonami's bespoke practice is evident during our online interview. Various hardware components are strewn across the workbench in her studio, alongside a home-made analogue synthesizer in preparation for a collaborative online performance the next day, showing the fluidity and spontaneity of her instrument-making process. In recent years, Sonami began experimenting with a range of materials and software, combining appropriated metal objects, contact microphones and interactive machine learning (ML) to create the Spring Spyre. Insights from her design and performance experiences demonstrate the depth and complexity of combining new interface design and movement-based performance, usually starting with a simple idea that develops through practice.

Sonami, like many of the other artists I spoke with, was initially attracted to the field to counteract the tyranny of the screen when making music with computers on stage. The concept of a glove instrument arrived as an alternative to diverting eye contact and attention away from the performance and audience, Sonami recalls during a recorded Zoom conversation in July 2020, which is the source of all interview quotes in this chapter:

> I started using computers, and then I thought what am I going to do? I can't stare at the screen. I was talking to my friend, Paul DeMarinis. He's actually the one

that showed me, cause at the time you didn't have Arduino. It was 1991. I didn't know how to program microcontrollers. He said here's a system, and I thought, well I'll just use a kitchen glove. It turned out to be funny. But actually there was something I think about making a gesture and hearing a sound at the same time that just blew me away. Do you know the magic of it?

Once Sonami experienced this new connection between gesture and sound, she felt instantly transformed by the possibility of channelling her movements into sound, video, lighting and vocal effect controls. By attaching five hall effect transducers that measure magnetic field strength to her fingertips and a magnet to her right hand, Sonami was able to generate a signal based on the proximity data of the fingers to the magnet to drive an assemblage of synthesizers and samplers using musical instrument digital interface (MIDI) protocol, an industry standard developed in 1982 to connect micro-processor-controlled hardware produced by different manufacturers. Wishing to move without spatial reference by not having to position herself in relation to a sensor, Sonami chose a wearable approach, initially creating gestures inspired by Indian dance and sign language. She also desired multiple, parallel controls, changing the mapping and sonic material for each composition. Combining samples with abstract sounds created using frequency modulation (FM) and additive synthesis, Sonami later added voice, articulating the cinematographic texts of writer Melody Sumner Canahan to present a musical form of storytelling. During the interview Sonami explains how her evolution with the self-made instrument deepened with each performance:

> At the beginning it was tongue and cheek but then I think it was one of those lucky things where I grew with it. I grew to understand the relationship, and I don't think it had any political bent at first but maybe as it grew, this whole idea of clarity, and control, and entertainment became a lot clearer to me. Am I your entertainer? Am I my entertainer? Who am I entertaining? I think about the spectacle, that I had to look good. I still had to look convincing.

Themes of transformation, spectacle and entertainment expose the theatricality of featuring gestures within performance. Highlighting gestures raises aesthetic considerations, drawing attention to the performer's body and style of moving. Compared to instrumental and laptop performance, which elevate the utility of objects as supportive props, audience expectations are projected directly onto the performer's body when the instrument is largely invisible.

Sonami exploited these aesthetic possibilities by experimenting with a range of materials in subsequent versions of the glove. The rubber kitchen glove of the first prototype strips away seriousness with implied domesticity, linking live electronic music with the mundanity of washing dishes. The instrument assumes satirical and humorous overtones, contrasting with the bulky masculine apparel featured in virtual reality systems like the monochromatic Mattel power glove. Sonami's wearable creation evolved from its initial crude rubberized form into a slick cyborg glove with exposed wires resembling nerves emerging from the tendons, skin and muscles of the hand. Bert Bongers (2000) built this later version in 1994 at the Studio for Electro-Instrumental Music (STEIM). A range of sensors are sewn onto black mesh including blue bend sensors for three fingers, microswitches on the fingertips, ultrasound sensors and an accelerometer. His next iteration of the glove in 2001 featured metallic bend sensors visible through transparent shrink wrap. The final version in Figure 1 has two accelerometers on the right wrist band that detect speed and motion, an ultrasound receiver for measuring the distance between both hands and a finger band holding a miniature microphone.

The signals are channelled through the STEIM analogue to MIDI interface, SensorLab, designed to connect physical phenomena like gestures to personal computers and studio devices using MIDI protocol. The data is streamed to computer

Figure 1 Laetitia Sonami and the lady's glove. Photo credit: Frank Baldé

software, Max/MSP (n.d.), where movement to sound mappings are programmed in advance of performances. Although initially designed as a gestural controller, the lady's glove later evolved into an exploratory system, encouraging Sonami to 'move beyond illusions OF control' (Sonami 2014). During peak immersion, Sonami became enamoured with the instrument as a liberating force: 'The lady's glove, I could still just cry thinking about it. It's a beautiful instrument. Sometimes it's just like flying.' This floating feeling captures a physical process without restriction and resistance, almost transcending the technology that created it.

Sonami's performances with the lady's glove sought to establish a reciprocal understanding between the performer and audience to compensate for the lack of familiar and predictable relationships between movement and sound that characterize traditional instrumental performance. Creating works that could be understood or at least have some impact on the audience became a primary consideration when pre-planning mappings between synthesized and found sounds, live vocal input and gestures, Sonami reflects:

> Part of it was like my body discovered a new way of performing. I was also learning things that didn't work. If it's too complicated, then it affects the whole relationship with the audience. Do they understand what I'm doing? Am I supposed to be their little slave girl that's going to lift an arm to make a sound? It's not just this relationship to myself, but the relationship to the audience.

The notion of the gestural performer as a muse and entertainer centre stage, with all eyes on them, is disconcerting for Sonami. She reports a sense of envy of laptop performers, who can largely ignore the audience, whereas gestural musicians are not afforded this luxury, leading her to feel like a 'glamorous waitress' (Sonami, cited in Rogers 2010, 229) fulfilling the traditional service role of the entertainer offering up a visual spectacle for the audience. This theatricality in motion-based performance can evoke feelings of discomfort, particularly for women, who are more likely to be judged for their appearance than their male counterparts. The sounds produced become almost secondary to the external image and charisma of the performer.

When playing the lady's glove, Sonami's physical expression was not limited by preconceived gestures developed specifically for the instrument. Her movement language evolved with the open and flexible design, allowing Sonami to realize a new performative identity through technology of her own invention:

> It really changed my identity in the sense that what I learned through it was that I cared about finding ways to create an instrument or create a system that

would allow me to be different. Why do we not build more instruments that change us? Don't tell me about performance with Ableton. I know there're ways to do it, but when you're on a matrix of 108 buttons, you press the button and the damned thing happens. There's nothing more satisfying. It's like crack. So I think about the whole idea of change and transformation, and then I'm not political but if I had the desire at the time, I would think more about how we create instruments for change.

The unique glove design inspired Sonami to move in unexpected ways, leading to a sustained period of design development and long-term engagement: 'It made me do things in a certain way that I would never have done if I had a keyboard, so that was a surprise. So I bend one finger and something else happens. That's interesting. That was not of my own volition,' Sonami observes. The instrument thus allowed her to transcend the clearly delineated parameters of touch-based commercial controllers connected to performance-based music making software, Ableton Live, in which samples and song sections can be triggered reliably and consistently according to a set tempo. Sonami's interaction with her instrument extends beyond issuing musical commands with minute button presses. She prefers to work cooperatively with the lady's glove to reach unpredictable musical outcomes, speculating:

The instrument, in a way, becomes a way to go beyond. The thing is, if I'm thinking about the instrument as control, I won't be able to land in expected places. So, it's like guiding it. A lot of it is a lure. I'm very aware that I'm trying to pretend I landed somewhere new, when actually I probably just landed around the corner, but there's still the hope that you hear a sound that you've never heard.

In her *Pink Noises* interview with Tara Rogers (2010), Sonami explains how she pre-plans sounds and mappings before her solo performances, finding it challenging to maintain a sense of fluidity when improvising with predesigned materials. She addresses this limitation by applying templates that contain inbuilt possibilities linking certain sounds, filters and resonators to specific sensors that can be dynamically remapped.

Eventually, Sonami retired the glove once the interaction began to feel predictable. The instrument had satisfied its original purpose of creative renewal and growth. In our interview she reiterated the need to incorporate more flexibility:

I never had to witness the glove. I never had to witness my body. But also, the lady's glove did not really surprise me that much in the way that I programmed it at least, cause a lot of it was in the mapping. It's all in the mapping. I didn't map it in a way that would behave in an unexpected way.

Having reached the limits of the self-designed lady's glove, it appeared that the performer and instrument had blended and become one. In her requiem for the glove, Sonami states that it had melded with the muscles and tissues of her left arm, becoming part of her, immediately operable without forethought:

> I stopped playing the lady's glove because my imagination and the lady's glove had merged into one. I knew what it could do, and I knew what I could do… Gestures, sounds and geographies slowly became mummified.
>
> I left it behind because I did not want to blemish it with poor looking associations. So, I leave gloves and other wearables to small dictators and corporate powers. (Sonami 2017)

Questioning the growing commercialization of wearable sensors, Sonami felt the need to pursue a more individual path, shunning the notion of functional and predictable glove controllers with their 'parametric monotony' (Sonami 2017, 140) linking singular gestures with predictable sonic outcomes like bending a finger to alter vocal pitch. Sonami also foresaw the inherent dangers of burgeoning personal biometric products. The increase in personal fitness trackers prompted her to relegate wearables to the realm of mass surveillance tools that syphon off individuals' physical habits and data. These devices, touted for promoting well-being, measure and collate statistics on individual health, from the number of steps walked daily, to oxygen levels in the blood and sleep quality, quantifying user experience.

Observing the potentially negative impact on users' privacy, Sonami firmly rejected the capture and transmission of intimate body metrics to corporate data farms for processing, analysis and targeted marketing. She discarded the lady's glove once it no longer served the process of personal transformation and expression and the technology behind it was appropriated by commercial forces. Another reason for retiring the glove was the pressure of being typecast as a performer of a single self-designed instrument. In a field that values continuous innovation and improvement, artists can be judged for continuing to play the same instrument for too long and even viewed as bereft of ideas unless they demonstrate continuous technical innovation. Customized systems for one-off performances are prevalent in conferences and concerts featuring gestural systems, indicating the high value placed on new designs.

Towards the end of her time performing with the lady's glove, at her 2014 NIME keynote, 'Dreams of Control, Dreams of Chaos', Sonami departed from her scholarly presentation as her voice was disrupted by technical interference, dissolving into static and a cyborg-like, processed vocal line with the aid of hand gestures sensed using the elbow-length black mesh glove on her left hand. She disassembled the conventions of academic speaking in one fell swoop, enacting a vocal disruption by telling a story through abstraction, 'moving through archaeological layers of meaning from the voice', before abandoning the voice altogether. A combination of sweeping, stretching arm movements and intricate hand gestures initiated jet take-off recordings, song material and spoken word samples. With slow, deliberate arm movements and sculpting motions, Sonami layered synthesized sonic textures, scrubbing through, elongating and rearranging arpeggiated and resonating phrases by balancing thirty sensors while maintaining an air of effortlessness. The dissolution of voice into purely electronic tones represented a shift in Sonami's later performances with the instrument away from featuring vocal expression once she realized 'at some point I had to stop because the voice and the stories became too important. The sounds became poor, cause it was easy for me to do the voice. Then all the things I had done had these interesting stories, and then I thought, the sounds are getting weak. So, I went OK, no more stories, no more voice, and then I stopped.' Temporarily reaching the limits of her vocal delivery, Sonami entered into new instrument design once again, gathering materials and ideas for her next instrument.

Sonami incorporated several elements from the lady's glove into the Spring Spyre, which evolved without intention, almost serendipitously. Named by composer Eliane Radigue, who composed the first piece for it, *OCCAM IX*, it represents a move away from spectacle, exploring elements of randomness. Sonami is less concerned this time about making the links between gesture and sound clearly visible and transparent to the audience. Three thin springs are connected to three audio pickups fixed to a recycled steel ring, as seen in Figure 2. Unlike the lady's glove, where 'hand movement is tightly mapped to sound' (Fiebrink & Sonami 2020, 239) the Spring Spyre inspires improvisation and unpredictability, fed by three audio signals produced by touching the strings that are then analysed and sent as inputs or features to drive ML models trained with Rebecca Fiebrink's Wekinator software. The Wekinator uses these models to control several synthesis methods in real time through Max/MSP. A Roland PC1600 controller facilitates mixing of sound synthesis. The level of randomness

Figure 2 Laetitia Sonami and the Spring Spyre. Photo credit: Brown U

can be changed based on how 'narrow' the training set is. Wider examples lead to more unpredictable results according to Sonami:

> If I feed the system training examples whose sounds encompass wide changes based on how I touch the springs, the trained models will move through all these points in unpredictable ways as the springs settle to a resting place. If I give it training examples with narrower changes, the sound will just oscillate slightly as I move the springs. I can thus easily scale the instrument between predictable and unpredictable results by changing how I train. I refer to these variations as the 'synthesis terrain', a nod to David Wessel's 'Timbre space'. This 'predictability index' is very easily modified and unique to ML. (Fiebrink & Sonami 2020, 239)

The instrument developed over several years, emerging like a baby for Sonami, who often wonders, 'Where do you come from?' They are still to get to know each other, as the instrument progresses through infancy, adolescence and then adulthood, 'trying to be kind of a direct transmission with a system that you're part of' Sonami ponders:

The lady's glove definitely had that feeling that it was under my skin. I was just part of it, and I'm trying to understand why the Spring Spyre, like I wrote in the requiem for the lady's glove, is outside of me. It's not in me. It doesn't occupy me. The relationship is very different, like a foreigner, something I have to understand. The lady's glove I never had to understand. Strange, isn't it?

Sonami doesn't regard the Spring Spyre as an instrument, but rather as a system with a unique personality underpinned by its behaviour and material. Unlike the intangibility of the lady's glove, its characteristics are shaped by Sonami's commitment to training and retraining the ML models that bridge the audio signals produced by her movements in relation to prominent metallic hardware, and sound synthesis. Sonami compares it with earlier self-made synthesizers, which never recreate the same sound twice.

After firstly consolidating the Spring Spyre's structure, Sonami then concentrated her compositional efforts on refining the models for controlling sound synthesis. ML offered scope for musical experimentation, acting 'as an open system which allows for continuous exploration in sound synthesis, expressivity and adaptability' (Fiebrink & Sonami 2020, 239). Mapping becomes a link between the physical and sonic world: 'ML offers a way to easily configure mappings with a wide variety of behaviours, thus allowing the composer to focus on the sounds and compositions' (Fiebrink & Sonami 2020, 240). Sonami's fruitful collaboration with programmer Fiebrink was built on joint hacking sessions, accenting the role of creative collaborations between artists and software developers in the direction and functionality of code used to craft new instruments, and culminating in a co-authored NIME paper that addresses a gap in ML literature which overlooks long-term instrument building and performance applications. Sonami regularly provided feedback on paper prototypes and suggested features she would like to see as the Wekinator software developed. This artistic input had a notable influence on the code. Fiebrink also conducted participatory workshops and interviews with composers to feed into the iterative software design process, finding that it encouraged play and exploration of gesture to sound relationships rather than focusing attention solely on the underlying programming and mathematics of the Wekinator. User feedback from the workshops revealed that the software 'made instrument design accessible to non-programmers' (Fiebrink & Sonami 2020, 238). Yet based on the observations of intensive, long-term users like Sonami, Fiebrink was able to determine in a more targeted manner which features to prioritize and improve for performers.

Sonami identifies one of the major challenges of ML as the inability of current synthesis methods to handle vast numbers of parameters simultaneously (Fiebrink & Sonami 2020, 240). This number can be as large as eighty inputs in Sonami's works. Yet she rails against the accurate and predictable control of individual parameters in prepared pieces:

> I think most often musicians apply control and mapping strategies after they have composed their pieces and resist opening up their compositions to allow these to interfere, influence or even hijack their original plan. This seems to be a poor application of ML but the desire for control still prevails over the desire to explore new forms of expressivity. (Fiebrink & Sonami 2020, 240)

Sonami's penchant for exploration and risk-taking is evident in a series of online quarantine concerts for the Experimental Electronic Music Studio in 2020. Sonami collaborated with a harpist and visual artist, presenting both instrumental and voice-based works. The first performance with Paul DeMarinis and visual artist SUE-C at the Quarantine Concerts is introduced with murmured vocal utterances and tongue clicks with effects, issuing a direct and intimate communication with the distant audience, before dissolving into abstract analogue synthesizer swoops and arpeggios. In the second instrumental performance, her entire upper body is framed by the circular Spring Spyre, as she caresses and gently tweaks the strings in an almost meditative manner. Sonami pauses to sniff an artificial flower mounted on the instrument. A hardware unit behind the instrument captures and processes sounds. There is an inner intensity, as if she has almost transcended the remote, invisible audience. These latest performances show Sonami at one with her instrument, a product of her imagination, embellished by conversations and idea exchanges with collaborators. Her involvement in the technical design of her individual instruments and systems brings new ideas to life in her live work.

Illusions of control

Sprung from satire, the lady settled into an indispensable interface, inextricable from my imagination. What first started as dreams of control and virtuosic display, turned into dreams of chaos.

(Sonami 2017, 139)

A recurring theme in Sonami's writings and interview is concerned with the distinction between controller and instrument. She is interested in a continuous, unfettered dialogue with her instruments, not limited to pressing buttons and triggering sounds through a controller that links specific actions to predictable musical outcomes. Sonami views her interaction with the lady's glove as akin to an intimate relationship: 'It's kind of like this strange reflection of many states of mind and it's long. It's 25 years you know. It's like my secret lover.' Both of her unique instruments helped Sonami realize and express new types of music unavailable through traditional instruments or instrument-inspired controllers:

> The musical instrument is a vehicle for the desire to make music. It is both something that must be internalised, incorporated and made flesh and something other, without which we could not get to that sound from the world beyond. (Ryan, cited in Andersen & Gibson 2017, 37)

Similarly, researcher in practice-led methods, Linda Candy (2020, 197) refers to the possibility of 'digital as partner'. Candy (2020, 181) observes that common terms in the field such as 'tool' and 'medium' are gradually being replaced by 'mediator', 'partner' and 'dialogue'. She references Joseph Lidlicker, who in the 1960s predicted a future time in which the human brain would become closely intertwined with the computer. In examining human–machine partnerships, Candy asks the following questions:

> [F]rom the human point of view, does being partners imply that there must be agency on both sides? Does a partnership require a demonstration of autonomy in thought and action? Is it enough to think of a partner as the other half of a duo engaged in the same activity?" (Candy 2020, 198)

This line of enquiry is particularly relevant to digital systems that incorporate artificial intelligence and ML, leading Candy to enquire whether fully automated machines can demonstrate equivalent agency to humans. Researcher, Andrew Johnston (2009) created a performance system for augmenting conventional instruments that explores these concepts. The system facilitates three interaction modes: instrumental, ornamental and conversational. When engaging in conversational interaction, the performer is able to relinquish complete control, forming a partnership with the system.

In a novel perspective on control known as dynamic coupling, O'Modhrain and Gillespie (2018) argue that it is not the musician who controls an instrument, but rather the dynamical system between instrument and the performer's body that governs the quality of the interaction: 'The musician is not playing the musical instrument but instead playing the coupled dynamics of his or her own body and instrument' (O'Modhrain & Gillespie 2018, 21). Their theory explains how a human–machine interface can become an extension of the body so that its operation fades from consciousness. The concept of dynamic coupling can be applied to a musician's sense of control of an instrument: 'An ideal musical instrument is a machine that extends the human body' (O'Modhrain & Gillespie 2018, 21). The mechanical system of the instrument interacts with the biomechanical system of the body in a symbiotic relationship. The resulting feedback loop incorporates force and velocity as the performer applies energy to the instrument.

Composer, performer and instrument designer Marianthi Papalexandri-Alexandri (Maier & Papalexandri-Alexandri 2020, 449), whose inventions possess their own agency, also explores an interdependent notion of control in her electroacoustic instrument-building processes, 'creating a completely dynamic relationship between sound, instrument, and performance, a constant process of gaining control and losing control'. Rather than approaching a piece with set intentions in a traditional compositional sense, Papalexandri-Alexandri allows the instrument to reveal its novel character and influence the formation of ideas and flow of a performance.

Yet there are limits to this two-way exchange. As mentioned in Part One, not all musicians are skilled luthiers or programmers able to conceptualize and customize their own devices to their individual artistic and functional purposes:

> Being able to customise one's personal devices is possible but usually at a relatively surface level. Digging deep into the software and hardware is a skill that only a minority possess. This means that the extent to which practitioners can control the technology to suit their needs is often limited. (Candy 2020, 181)

A lack of programming skills can affect the level of control artists exercise in their creations. Limited coding knowledge leads many musicians to collaborate with software developers to realize their full vision in digital performance, sometimes losing themselves or diluting their ideas in the process: 'That sense of control over the technology is different if people are able to design and construct the tools for themselves. For many creative practitioners, this is the preferred path because their ambitions for their artworks cannot always be satisfied using ready-made systems' (Candy 2020, 181). Sonami rued the frustrations of programming when she reached the limits of her knowledge, reflecting:

I wish I had a programmer who could do things much better than me. So much of that process is a craft. Learning how to craft this, learning how to do this programming, and soldering this thing, and that connection with the material is essential.

Yet working with Fiebrink to configure gesture-sound mappings through ML allowed Sonami to shift her attention towards exploration and play rather than focusing primarily on programming and code during the conceptualization and development of the Spring Spyre. This experience revealed the benefit of collaborating with expert programmers to realize more sophisticated designs. The shared process matches with Sonami's tendency to slide between varying degrees of unpredictability and control rather than pursuing a strict control paradigm (Fiebrink & Sonami 2020, 241).

Control is also compromised by the dynamic nature of gestural systems with constantly changing features. Renowned gestural musician Michel Waisvisz had a disappointing experience when first performing with his self-designed instrument *The Hands* in 1984 (Torre & Andersen 2017, 132). Much of his discontent could be traced back to cognitive overload caused by managing multiple processes simultaneously. These processes include manipulating a range of sound sources and his own voice with small keyboards worn on the hands, which incorporate force and tilt sensors to sense hand inclination. His fingers controlled the keys, while the thumb operated a pressure sensor, and ultrasound transducers measured the distance between the hands. The movement information collected by the sensors could be mapped to diverse sound parameters, from pitch to loudness and timbre (Bongers 1998). Following the frustrating performance, Waisvisz took a brief hiatus from performing with *The Hands* to focus on instrumental development, using his discoveries as a template for artist-driven technology workshops at STEIM.

Waisvisz froze development of *The Hands* in order to perfect performance techniques for live composition. The temptation to keep tinkering with the interface, which many performers engage in the night or even an hour before a performance, sets up a technically risky situation in which the performer feels unsure how to handle their newly refined beast. They hang by the seat of their pants, uncovering new idiosyncrasies and vulnerabilities during public performance, as Waisvisz notes about his gestural performance practice:

About my own experiences with gestural controllers I can only say that I fight with them most of the time. That's something that almost every instrumentalist will tell. But if you are in the position to be able to design and build your own instruments, and so many interesting technologies pop up almost weekly, you are tempted to change/improve your instrument all the time. This adds another conflict: you never get to master your instrument perfectly even though the instrument gets better (?) all the time. The only solution that worked for me is to freeze tech development for a period of sometimes nearly two years, and then exclusively compose, perform and explore/exploit its limits. (Waisvisz 2000, 629)

Balancing technical and functional imperatives with artistic considerations can be achieved by temporarily suspending system updates: 'By freezing the design modifications and extensions when The Hands were physically stable and durable, it became possible to focus on the musical intent beyond the novelty of the devices and engage in the aesthetic and musical considerations' (Torre & Andersen 2017, 134). This development, nurtured by practice and use in different contexts, consolidates and extends the features of novel gestural instruments periodically. Torre and Andersen (2017) classify this as a customization phase in which the artist develops the physical skills to engage in continuous movement, realizing predesigned mappings that bring forth sound. This process allowed Waisvisz to invent novel movements in real time, responding to and interacting with the sounds he produced:

And I think that if you would analyse all the movements of what I do, it's interesting because my instrument doesn't require that much specific movement, but I would easily guess that more than half of the movement is to connect and to help to shape time, rather than it being functional to the triggering like interacting with the sensor or so to say. (Waisvisz, cited in Torre & Andersen 2017, 134)

Sonami adopts a similar method of immersing herself within a singular system and learning during instrumental development and performance about the types of movements that are most conducive to shaping sonic material over time:

For me, there are some gestures that are really relevant. So much has to be extreme concentration with 30 sensors. The 30 sensors are mapped to do certain things. I have a kind of story. I have to pay attention to my body but a lot more is like meditation, emptying and relaxation. It is not just about the body or the music, but about this space that I'm trying to go to, and I have to really focus. I don't know what that space is. I just know that it's there.

Sonami creates interaction spaces that evolve through practice. Inward focus and intense immersion during performance allows pre-planned gesture-sound mappings to mature, solidifying connections between performer movement and sonic processes:

> Sounds, events, emotions migrated to specific gestures, and some became fused:
>
> The bending of the wrist (upwards / palm down) attached itself to mechanical sounds, functional or dysfunctional (pumping, breathing machines and various clicks);
>
> Hands moving back and forth became changes in density or changes in acceleration as they scanned through layers of sound deposits;
>
> The bending and scribbling of all left fingers like a spider's legs, to speech utterances or small events nearby. (Sonami 2017, 139)

This long-term engagement with the instrument enables the development of a personal expressive language and virtuosic skills (Torre & Andersen 2017, 135). It is this practice that guarantees the system's instrumental status: 'It seems that the time invested and the artists' perseverance may be a key to an instrumentality that is achieved by favouring long-lasting ageing processes over a habit of fast-prototyping and fast-dismissal,' Torre and Andersen (2017, 135) argue. This latter rapid development, unless deliberately frozen, differs substantially from gradual acoustic instrument evolution which has resulted in instruments barely changed since the nineteenth century (Magnusson 2019, 60). In contrast, a gestural instrument or system design is rarely complete – it evolves in parallel with the performer's sensorimotor skill development.

John Ferguson's (2016) analysis of Waisvisz's solo performance with *The Hands* at the No Backup concert, *HyperInstruments I*, addresses the challenges of controlling and asserting autonomy when performing with a customized instrument. The unpredictable and organic process is also evident in Waisvisz's performances with his other invention, the Sphinx. He likens the experience to struggling with a mythical creature, embracing the messiness of shifting performer–instrument relationships that challenge human autonomy on stage, offering creative possibilities on par with acoustic instruments (Ferguson 2016, 256).

Some practitioners in the broader interactive arts realm deliberately incorporate unpredictability into their systems, like Sarah Fdili Alaoui, whose work, *Double Skin, Double Mind*, employs physical simulation and particle systems. Rather than making a prescriptive system that responds to specific postures and movements, Fdili Alaoui invents one that draws out the movement qualities of the performer (Candy 2020, 189). The result is a partnership between digital technology and the performer in the work, rather than a one-way power relationship where the artist exercises strict control over a system.

In gestural system performance, musicians are continually challenging the myth of the skilled virtuosic performer as the ultimate pathway to control. Sonami and Waisvisz stress the need to suspend the tendency to feel in control all the time. Yet there is an inherent risk in expanding the traditional notion of control. Updated software and hardware systems often fail when not tested adequately before performance. Artists also occasionally experience discomfort as they engage with novel technologies and try new movements publicly for the first time. Musicians might be temporarily distracted due to an underlying technical problem arising from frequent system updates, unable to focus on their initial musical goals or perform with confidence.

Machine breakdowns and physical awkwardness on stage both contribute to a temporary loss of control. Prevailing metaphors of attempting to wrestle control from highly idiosyncratic set-ups include taming a wild beast (Ferguson 2013) and facing the dangers of bullfighting (Jordà 2004). To manage this precariousness, many musicians in the field create a framework or foundational structure for broad control then leave other parts to chance, like Sonami's Spring Spyre, where gesture–sound mappings can be recreated based on the quality of training data fed into ML software. Further challenging the usual conceptions of control, Rogers (2010, 7) reflects on the emphasis placed on precision, control and mastery within dominant techno-scientific narratives, particularly in areas like electronic music production. These established notions of control need to be widened to accommodate gestural performance practice, in which musicians grow and develop in line with their customized and imperfect machines. Sonami sees value in the 'awkwardness of an external apparatus, a mechanical system that the body's trying to adapt to, and the struggle that comes with it' (Sonami, cited in Rogers 2010, 229). Within this struggle Sonami recognizes opportunities for instruments to transform musicians' performative and compositional approaches.

4

Atau Tanaka

Prolific in documenting his practice, composer and researcher Atau Tanaka (2000) has dedicated substantial effort to exploring the intimacy of how the instrument 'feels' to the performer and the types of skill-sets musicians develop to control sensor-based instruments. This theme emerges strongly throughout our interview, as Tanaka outlines the design challenges associated with gesture capture, classification and mapping. Tanaka's compositions and performances feature no external instrumental object, but rather draw on internal physiological muscle-sensing data, typifying the notion of body as instrument. Tanaka has explored gestural interaction in solo and ensemble performance using the BioMuse as his main system for live computer performance over the past three decades. The intimate sensing system detects muscle signals through electromyographical (EMG) sensors attached to the arms. Tanaka's work is sonically driven, drawing on diverse influences from Japanese noise music to John Cage. His recent artistic explorations have integrated machine learning (ML) in mapping actions to sonic processes.

The BioMuse interface senses and digitizes neural and muscular activity. It is designed by researchers from the Centre for Computer Research in Music and Acoustics (CCRMA) at Stanford University, Hugh Lusted and Ben Knapp of BioControl Systems. Electrodes on the skin capture electrical signals from the muscles (electromyogram, EMG), brain (electroencephalogram, EEG) and heart (electrocardiogram, ECG) activity (Lusted & Knapp 1998). A digital signal processor interprets the data to control sound through a synthesizer using musical instrument digital interface (MIDI) protocol. The BioMuse captures continuous spatial gestures. MIDI data derived from the performer's muscle tension is mapped to various frequency modulation and waveshaping vector (WV) sound synthesis parameters. Tanaka was commissioned to compose the first musical work for the instrument, *Kagami*, which debuted in 1990. The BioMuse then became his main instrument.

A unique aspect of EMG data is that 'with the EMG signal we're picking up the neuron pulses of the body *preparing to* do the gesture, and so we're not picking up the results of an action, we're picking up an intention of an act', Tanaka explains during a recorded Zoom conversation in June 2020, from which all interview quotes are derived. This can make muscle-sensing information difficult to predict and therefore manage. There is also the problem of how to map a multitude of physiological signals to musical processes, Tanaka observes:

> The EMG signal – the muscle signal as a kind of interface for musical gesture – produces a signal that is much messier than a potentiometer, distance sensor, or a bend sensor. To make sense of that body signal, that physiological signal, it seemed to me that some kind of advanced information analysis like machine learning could be a really good case for it, and that hopefully machine learning could be apt to deal with this signal.

This realization prompted Tanaka to explore neural networks for processing EMG data more effectively. Neural networks are a collection of algorithms for recognising patterns within a set of data modelled after the interconnected neurons of the human brain. Like Sonami, Tanaka adopted Fiebrink's model of interactive ML, which allows users to add training sets and edit the training data using real-time, open source software, the Wekinator. The software facilitates gesture analysis, allowing users to build interactive systems by demonstrating gestures and computer responses without the need to write programming code. The Wekinator can be packaged into a Max/MSP patch called RapidMax, forming a modular addition to the performer's musical system environment. Although Tanaka first began experimenting with ML using a neural network object by University of California, Berkley, Centre for New Music and Audio Technologies (CNMAT) in the 2000s with David Wessel and the Wekinator from 2012, it was not until 2019 that he brought it on stage and experienced musically satisfying results.

Preparing the physiological data for the ML algorithm and conceptualizing mapping connections between movement and sound emerged as key challenges for Tanaka when applying the technique:

> The problem was to understand what you wanted to create out of what – so what was coming in in terms of signals from the body and what you imagined that to do in the music. That is ultimately a mapping problem. It's just a mapping problem that isn't a direct hard wiring mapping problem, so you don't get to

decide: I'm going to take this thing in and map it to this one parameter. This is like many parameters in to many parameters out.

Tanaka's long-term goal is to achieve a more sophisticated machine interpretation of the gestures reflected in the EMG signal. It involves taming the extensive data generated by the BioMuse by separating and contextualizing it effectively. Human limb gestures derived from EMG channels need to be segmented before they can be identified by the ML algorithm. The 2007 performance image in Figure 3, for example, depicts prototype armbands capturing muscle tension in the anterior and interior muscle groups that are placed on the forearm extensor muscles, from which a computer model calculates wrist rotation using simple pattern recognition.

In more recent performances, Tanaka aims to make the ML process visible and legible to the audience by training the model live on stage rather than arriving with a prepared algorithm and existing database of gestures:

> The process of performing the piece – the beginning part of the work is actually to provide the neural network with that training data live on stage in front of the audience and get it to train. These algorithms run fast enough that you can

Figure 3 Atau Tanaka at STEIM (Studio for Electro-Instrumental Music) Amsterdam. Photo credit: Frank Baldé

now get it to do this. Then the second section of the piece would be running the neural network and exploring that gesture sound space. This has got the added benefit of exposing the process of machine learning to the audience, which I hope is illustrative but also insightful to the audience. To walk on stage with this incredible pre-trained model, it's all whizz-bang and then you say it's machine learning, but they have to believe you. So that's the kind of thinking that I've gone through to come up with a satisfying gestural performance that uses machine learning.

The training process involves recording stationary poses, which Tanaka calls anchor points. The software interpolates between the postures to record a novel type of choreography tailored for each performance. This dynamic process allows Tanaka to represent and accommodate his fluctuating energy levels during a performance rather than limiting him to a specific gestural language as a pre-trained model would. It can also make the system relevant to other performers with different movement approaches and energetic input, who can each customize the instrument to their own bodies by training the model based on recordings of recent performance actions.

Only recently did Tanaka discover a workable solution for integrating ML in his performances, aided by the Gesture Variation Follower (GVF). The software was developed by Baptiste Caramiaux during a European Research Council project, Meta Gesture Music. Building on existing gesture recognition techniques, Caramiaux's (2014) GVF algorithm can identify a particular gesture while it is being performed and track gesture variations over time. The software estimates the change in scale and speed of a gesture in real time in order to analyse the variations that differentiate individual performers and gestural performances. It captures how gestures are performed, increasing the levels of nuance available through gesture recognition methods. This research eventually led Tanaka towards re-engaging with neural networks composed not only of gestures but also a library of different associations between physical movement and sound. He can then navigate between anchor points that act as 'reference points in the gesture to sound association and then once it's trained and running, the neural network gives you a live regression model that is a continuous information space that I'm able to explore with continuous gesture'. These points or static poses form an integral part of the composition for Tanaka:

What you base the algorithm on are different examples or different associations of human body movement and sound, training the regression algorithm on static poses, where I might find a pose here and say here is a posture, not a gesture but a posture, and I would like this posture to correspond to this sound, and then I'd set up a synthesis algorithm, tweak its parameters, get the sound that I like and say this sound corresponds to that pose and then I'd find another pose and program another sound and say I want this position to correspond to that sound, and then I'd do any number of poses and sounds, and then the neural network trains on that data and creates what we call a regression model that is an information space that maps those poses coming in to those sound synthesis parameters going out.

As part of his performance preparation, Tanaka preprograms sounds using a variety of synthesis methods, including concatenative and granular synthesis. He identifies grains to explore and also matches that sonic work with his unique gestural language:

I am literally walking on stage with a blank slate in terms of the neural network, but there is a composition, so I know which sounds I've programmed. If I'm using granular synthesis, or I'm using concatenative synthesis, I know which sound units or grains I'm interested in, and I have my gestural language. I know some of my basic moves.

Tanaka identifies several common gestures that characterize his personal movement language. Without an object to guide and orient his actions, he distinguishes between ballistic gestures that are goal-oriented and performed quickly, and slower non-ballistic or more static gestures. Either can make a significant difference to the patterns of muscle activation though they appear similar (Tanaka et al. 2019, 5).

Tanaka does not craft his gestures in a particular way, but rather moves like a musician in a manner that 'leverages our intuitive associations between sound and embodied movement' (Tanaka et al. 2019, 5) as he mentions during our conversation:

Musicians move, but usually aren't taught about how they think about how they move – they just move, and if it's intuitive, and it's natural, and it looks good, we're happy. Because as musicians we're focused on sound.

As a result, Tanaka's works are sonically rather than choreographically motivated. Like many instrumentalists who endeavour to leverage their acquired skills and performance techniques, he emphasizes the need to discover a unique movement vocabulary that draws on individual cultural, historical and personal preferences. Tanaka is particularly interested in capitalizing on the main types of gestures that musicians make:

> I'm interested in musical instrument gesture, but to get to that point, that gesture, that expressive ancillary gesture, and to make my whole musical universe come out of that. The musicality of musical instrument performance comes from those ancillary gestures, as Marcelo Wanderley talked about, that are not necessary for the physical, acoustical production of sound, but it is those extra gestures, the body English if you may, that allows us to articulate the musical phrasing. I was interested in getting to the essence of musical gesture in that way.

Godøy identifies ancillary gestures as those that musicians make spontaneously, 'shaping the performance on a higher level of motor control and musical intention' (Godøy 2011, 75). Tanaka's observation about their role in phrasing articulation is supported by numerous studies demonstrating the expressive role of ancillary performer gesture. Motion-tracking studies of pianists' and clarinettists' performances reveal that ancillary gestures like head and torso sway support and accompany sound producing gestures by conveying artistic intentions and visual cues to the audience (Wanderley 2002; Davidson 2012).

To classify and interpret these inherent variations between artists and performances, Tanaka has used Laban Movement Analysis (LMA) common in dance to understand movement flow and effort within peripersonal space or the kinesphere, the area surrounding the body measured by the full extension of the upper limbs (Laban & Lawrence 1974). The LMA method provides a qualitative framework for describing how effort is directed and performed, making it useful for collaborative projects to provide a shared language for contributors across different disciplines. Qualities of swiftness or pace can be applied to analysing the unique movement style of a performer. Tanaka finds LMA particularly relevant to muscle sensing, which does not fit neatly into spatial or Cartesian descriptors:

> If you're describing gesture either in some kind of graphical form like Laban notation or taking motion capture data and saying this gesture will be out front 90 degrees and then off to the right 45 degrees, then you're describing gesture in Cartesian terms that are spatially driven, but there's a lot in muscle gesture that's not spatial in that Cartesian sense and each person might exert their muscles in different ways. So I was looking into LMA to find a way to describe muscle gesture not in Cartesian terms, not in pure terms, because all bodies are

different, but to accommodate the diversity of bodies, to then think could we describe them as a function of Laban's kinesphere, effort and flow and so forth?

Tanaka recognizes a clear connection between LMA and soma-based design methods that are guided by bodily experience, seeing it as a more appropriate way to understand the subtlety and nuance of movement beyond measuring movement trajectories, or shapes traced in space (Ward et al. 2016). He contributed to a study that exposed the usefulness of LMA in aiding systematic movement exploration and authoring gestures for new digital musical instruments (Ward et al. 2016). Similar approaches are common in dance- and computer-based research, exploring the challenges of observing movement experience as part of a somatic design strategy (Fdili Alaoui et al. 2015). Fdili Alaoui developed a toolkit with EMG, proximity and inertial measurement unit (IMU) sensors, applying her finely tuned kinaesthetic empathy and skills in computational creativity to explore ML: 'I use the somatic practice, which is an additional set of tools for observation, to be attuned to one another, to train your kinaesthetic empathy, to listen to what's going on in the body and how you observe it, how you make sense of it, how you translate it' (Fdili Alaoui, cited in Candy 2020, 216). Both Tanaka's and Fdili Alaoui's analytical methods draw on LMA vocabulary to guide autoethnographic reflection on the role of movement in composition and performance.

Building on LMA insights into effort, Tanaka proposes the application of intention, effort and resistance to understanding and expanding performer relationships with the intangible BioMuse instrument and maximizing the creative potential of the EMG signal. In addition to LMA, Tanaka refines his sense of feel through Chinese martial art, tai chi and yoga practice. Regular yoga sessions outside of performance help Tanaka to direct his energy when manipulating invisible effort and inbuilt flows of resistance detected through muscle-sensing methods. Although it is not directly related to his performance work, yoga contributes to Tanaka's overall body awareness:

> Yoga is a very inwardly focused way of thinking about movement. It's not really thinking about movement, it's thinking about energy. All these things line up for me, in that gesture sensing that I use – the muscle sensors – and the way that I think about it is not just gesture as gesticulation per se in space. It is very concentrated, inward energies of effort and resistance. ... It's part of a daily or weekly practice to get in touch with my body and be aware of the world around me, and that's one of the most important things on stage – is establishing that

relationship between performer and the concert hall as an environment and inhabiting that space.

This physical practice emphasizes intention over the external manifestation of physical movement. It relates to the inward nature of muscle sensing. The body produces its own energy and its own resistance in the absence of an object to interact with. Tanaka (2015) explores the richness of muscle signals by identifying three modes of play – intention, effort and restraint – which the performer modifies in relationship to an instrument. The EMG signal captures the intention preceding limb movement, as it detects the neuron pulses of the body preparing to perform a gesture, unlike external sensing systems that pick up the artefact of the actual gesture, Tanaka points out:

> One interesting thing about EMG sensing is that there is no external object. The object is the body. Or the body is not really an object, if we go past the Cartesian mind body dualism, then we go past object/subject dualism as well. In any case, there is no external, physical material object to manipulate, and quite early on I thought about this and coined the term, boundary object, and this has a lot to do with effort. If you've got something you're working on – a wall or a ball that you're squeezing – that's a boundary object that provides resistance that gives you a sense of effort and something to direct the gesture to. This idea of boundary object goes away when using EMG. What I found about muscle sensing is that despite there being no boundary object to do the gesture to, that effort, and certainly fatigue, in muscle tension, definitely came around. Then the body, in producing its own resistance, can create effort. Coming back to yoga, that's why it's quite interesting to understand energy, to understand exertion of the body, so EMG as a sensing signal then, is the intention of that gesture.

When there is no physical boundary object to direct effort towards as with EMG-based muscle sensors, haptic feedback can assist in creating implied resistance. A feedback channel to guide gestural control and the direction of effort is beneficial when there is no central object to focus a musician's actions:

> In order to re-introduce a kind of resistance in sound and to create a kind of boundary 'object' to gesture, we can introduce haptic feedback through the physicalisation of sound output. Projecting audio of specific frequencies at amplitudes sufficient to create sympathetic resonance with non-cochlear parts of the body creates a haptic feedback loop through acoustical space of the effort engaged in musical gesture. (Tanaka 2015, 299)

Tanaka exploits the physiological foundations of auditory perception in mapping gesture to sound synthesis to create a haptic feedback loop for the performer that is internally sourced through the body, arguing that 'if muscle tension in the forearm can modulate the frequency and intensity of sound that is reproduced at sufficient amplitude to cause resonant vibration in the bone under that muscle, we have created a form of haptic feedback in the absence of traditional physical objects' (Tanaka 2012, 164).

This focus on physicality in performance stems from experience as a trained pianist and electric guitarist playing free improv, punk and surf music. Performing in festivals with Japanese noise musician, Merzbow – Masami Akita – also exercised a substantial influence on Tanaka's creative practice. His primary motivation is to address the challenge of incorporating physicality in computer-based music:

> As musicians, regardless of what instrument we play, we do use our bodies. What I was interested in was to preserve the satisfaction of what I got playing electric guitar, playing the piano, and transpose it to a computer, where we weren't supposed to have that sense of agency or viscerality. I guess my initial mission was to make the computer as satisfying to work on as a musical platform, as an acoustic instrument.

Since using gestural systems, Tanaka observed a definite effect on his identity, self-perception and agency as a musician, explaining 'my musical identity as a composer and performer, trying to find in digital systems a kind of an analogue continuity and to find the expressive potential in this, I had some inkling of some vision when I first started and it had to be imagined because things weren't as readily available as back then'. Tanaka connects a sense of agency with feeling good as a performer and draws on that feeling to relate to the audience. In this way, personal agency can then be shared with the audience:

> I think agency is incredibly important, especially in interactions with digital systems and the increasing complexity of these systems, whether they're on networks or multi-user participatory, or whether they involve layers of machine learning, or levels of indirection, which mean that agency can become impinged in the same way that we might not get a feel for what we're doing. Somehow to maintain that simple, direct sense of agency within the complexity of the system is of vital importance for me. I think if it feels good for me, it should feel good for the audience. For it to feel good for me, I have to have a sense of what I said I'm doing, so a sense of agency is primordial for myself as a performer to give

a good performance, and if I have a sense of agency in performance, I think then the audience gets a sense of agency, whether they are performing or not. … They are engaged and there is agency in the performer, and I think through mechanisms of mirror neurons, or just a sense of intersubjectivity, agency can then be transmitted.

Enhanced personal agency, stemming from how good a system feels to the performer, raises musical satisfaction, according to Tanaka:

It's got to feel good, and certainly with the muscle sensors, it's very much about how the musical satisfaction comes from how good it feels. And with machine learning it becomes difficult because this is an added layer of abstraction that could feel very interesting cause it's not hardwired one-to-one. Then this layer of indirection, you might call it, can make it quite risky in that it doesn't feel quite direct.

Yet the benefits of ML techniques, which address the complexity of mapping sound to gesture by learning rules based on examples eliminate the need for manual programming, potentially outweighing these risks (Visi & Tanaka 2021).

Interactive ML features in Tanaka's piece, *Wais* (2019). In it he uses Thalmic Labs' Myo armband to record electrical muscle impulses, as shown in Figure 4.

Figure 4 Atau Tanaka performance of *Myogram* (2015). Photo credit: Martin Delaney

The structure of the piece includes an opening section for auditioning samples from a Waisvisz performance recording featuring *The Hands*. A second section involves training a neural network, and the final section navigates through the trained model. Tanaka was impressed by the efficiency and sensitivity of the Myo band but considered it one key component of a larger system:

> I was amazed by how good the Myo was, and the fact that it was just there as a product meant that anyone could do it. But then it still means not anyone can do it. You still need the music. You need a way to map the gesture to sound, and you need the performance practice. It became interesting to me to work out how to transmit this performance practice that I have with muscle sensing. I'm certainly not the only one. That's where other artists like Marco Donnarumma, who studied with me, brought their own practice, and then having the related and different practices of a range of artists working with muscle sensing – myself, Marco Donnarumma, Miguel Ortiz in Belfast, and Federico Visi. We then begin to see how it might be meaningful for beginner users and young musicians. I became interested also in the EMG not only as a main instrument – I was interested in it for myself as a solo main instrument – but what if the EMG was used in conjunction with another instrument?

The Myo has since been discontinued. Despite capturing the attention of numerous researchers and developers, its one-size-fits-all approach made it less attractive to female users and it was uncomfortable if worn for long periods. However, as electronic circuit design has become more accessible, Tanaka contributed to the production of an open source myogram board as part of the Embodied AudioVisual Interaction (EAVI) group. Converting these discoveries to persuasive performances and enduring instruments is more challenging:

> Now we've got the systems to pick up the movement and turn that into music. For some musicians that might be interesting, but as we all know it's really difficult. It's not an easy thing to do. … The fact that it's research means that you can fail, so good if we try many things and a lot of things don't work. In any style of music, there are desires to do things, some of which are graceful and successful, and others that are valiant but it's really difficult to nail it.

Tanaka's involvement in research groups like EAVI and reflections on EMG sensing, interactive ML and recognition of intention as a key aspect of invisible muscle contractions provide valuable insights into touchless interaction. The *Meta-Gesture Music* (2017) album encompasses five years of research into new musical instruments that highlight the body, recording the visceral repertoire of key artists in the area. His solo performance for the compilation, *Myogram* (2015), shown in Figure 4, has a wavering intensity drawn from trembling forearm motions. Myo bands placed on each arm capture sixteen channels of EMG, translating muscle group activity during wrist rotation, finger movement and flicking gestures to sonify the raw muscle signal information. Processing in the form of low-pass filters, ring modulators, resonators and pitch shifters shapes the sound and controls musical intervals, directing it through an octaphonic sound system surrounding the audience.

Tanaka's EMG-sensing approaches also feature in an augmented piano work on the compilation with Sarah Nicholls, *Suspensions*, demonstrating their collaborative potential. By blending Nicholls's gestural approach to the piano, which integrates ancillary gestures in a highly expressive movement language, and Tanaka's expertize in producing gestures that optimize readings from muscle tension, the sonic character of the classical instrument is transformed through improvised, incidental gestures.

Some of Tanaka's discoveries with BioMuse and his own customized muscle-sensing performance system are also common to the next artist, Pamela Z, who combines this innate biosensing data with voice. She is featured in the next section exploring vocal and breath-based artists integrating gestural systems in performance, applying the notion of body as instrument within processed, sampled and looped vocal production.

5

Vocal and breath-based gestural systems

Like muscle activity, breath and voice are sources of data that emerge directly from the body. The voice resonates with the rest of the body during a musical experience (Eidsheim 2015, 164). It is heard as a vibration 'from within' (Merleau-Ponty 1968, 144), making it unique as an instrument because it is also felt directly through the body (De Souza 2017). The next two artists, Pamela Z and Julie Wilson-Bokowiec, combine extended vocal and operatic singing techniques with gestural control. Both performers portray a sense of theatricality and keen awareness of the impact of their physically based performance styles on an audience. Pamela Z merges muscle sensing with voice in her performances with the BodySynth, converting electrical impulses underpinning her limb movements into streams of data that control digital signal processing and looping of the voice. Wilson-Bokowiec uses gesture to explore inside her voice, splitting it into sonic grains that are then reassembled through live electronic processing. In a similar fashion, improviser Lauren Sarah Hayes samples and slices the voice into minute segments that are processed and reordered within dense textural arrangements. In the final part of the section, Stuart Favilla incorporates breath-based control of his intangible instrument, the LightHarp, while Joanne Cannon transforms woodwind breathing techniques and the sonic character of bassoon and contrabassoon meta-instruments within the Bent Leather Band.

The voice is considered the body's original and most intimate instrument (Overholt 2009), as the vocal sound emanates from the body, bearing the personal and emotional expressive imprint of the performer (Emmerson 2007). This view is reinforced by Don Ihde (2013, 103), who considers singing and other protomusical sounds like whistling, yodelling and throat singing to be the simplest and most physically expressive types of human-produced music. The voice, in partnership with movement, forms a potent channel of physical expression from infancy. Vocalizations facilitate connection with primary caregivers, ensuring that basic needs are met after birth. The vocal apparatus is hidden and reliant on autonomous functions like breathing, underscoring basic survival (Eidsheim 2015, 14–15). The inner workings of the voice are essential for survival yet mysterious and unknown:

> Voice, with its ability to filter out certain frequencies by changing shape, is often used to describe the synthesising process. Voice also plays multiple roles as an anatomical entity that protects the lungs from food and liquids, as a sound shaper, and as a transmitter of music and words. (Eidsheim 2015, 164)

The voice is thus an invisible acoustic instrument contained in the body, gaining visibility through facial expressions and body movements (Schloss 2003, 2).

Vocal production is explored throughout a range of disciplines including politics, phenomenology, linguistics, psychology and neuroscience. The pitch, pace and timbre of the voice conveys emotional cues linked to personal identity, relationship dynamics and worldly status. Each individual has a unique voiceprint, distinguished by the pattern, rhythm and sound of their vocal expression. This signature sound is one way a clan recognizes its members, reinforcing a sense of belonging within family and broader social groups.

The voice can inject uniqueness into performances, transmitting the vocalist's moods and physical endurance levels. This distinctive characteristic of vocal tone is harnessed by interaction designers to monitor well-being in an early release health tracker named the Halo Band by Amazon. The wearable device measures user energy levels and accompanying emotions through a small microphone on a wrist bracelet, analysing the emotional content of the voice based on pitch, intensity and tempo, while classifying high or low energy moments and positive or negative moods with subjective keywords like 'focused' and 'stern' (Fowler & Kelly 2020), offering an example of emerging speech-recognition applications that determine emotional states of users from vocal tone and patterns.

Vocal expression exhibits both individualistic and cultural aspects, making it 'physical, cultural, psychological and emotional, all at the same time' (Cook 2004, 10). Once the voice leaves the performer's body, it almost becomes a material object (De Souza 2017). This matter can then be manipulated through external means. In movement-based performance, once the voice emerges, it is possible to alter it through performer gestures, as Wilson-Bokowiec's performance works with the customized Bodycoder system and Z's BodySynth compositions attest. The vocal signal, altered through sensory data and processing, 'transcends the conventional body', expanding beyond its limits in a virtual, post-human era, as Martha Feldman (2015, 656) observes: 'It emerges from the body, inhabits it, invades it, overshoots it.' In live electronic music, the voice currently exists beyond physical boundaries, transported from the body through techniques such as ventriloquism or amplification, recording and processing.

The voice radiates a sense of belonging and contributes to meaning-making. Voice and speech are intrinsically linked, as apparent in the authorship of lyrical parts. Artists across disciplines are encouraged to 'find their voice'. Some vocalists achieve this by blending a range of influences like Bob Dylan (Feldman 2015, 664). The symbolic power of the voice distinguishes it as a highly flexible and multifaceted instrument, as speech language pathologist, Marina Gilman (2019, 62) notes:

> The human voice is the only musical instrument that has the ability to do it all. It can project over an orchestra. It is percussive and melodic. It can combine melody and text, and some forms of solo singing – known variously as overtone singing, throat singing, or harmonic singing – can even create harmony. And it is 100 percent portable; it goes wherever its owner goes.

The voice aids everyday communication and artistic expression. In the gesture studies field, where the term 'gesture' is used to denote hand motions and facial expressions that often accompany verbal expression (Kendon 2004), David McNeill (2005) regards gesture and speech as equally important in expressing thoughts. The close relationship between voice and movement in verbal communication, as well as vocal performance, has made gestural interfaces for the voice one of the most attractive and compelling subsets of gestural performance. From early pioneers, Michel Wasivisz and Laetitia Sonami, to Elena Jessop, Donna Hewitt and Imogen Heap, vocalists across genres are embracing gestural control of digital audio effects, sample triggering and looping. Gestural systems are ideally suited to vocal processing as they allow the performer to maintain eye contact with the audience while manipulating the timbre of the voice. Vocalists are able to harness their ancillary and expressive motions to augment and enhance a vocal performance.

Gilman (2019, 77) argues that the voice *is* the body. Vocal production is reliant on the life-sustaining cyclical inhalation and expulsion of air. The periodic rhythms of breath are regulated not only by the interior sensations of the musculature and organs of the body, but also by posture. Eidsheim considers vocal technique as reliant on 'modifying breath and the breathing process' (2015, 112). She recognizes voice, more than any other instrument, as reliant on a deep connection between the voluntary and involuntary processes of heart rate, hormone levels and respiratory function, as well as the inner feelings of the body. Breathing is the activity that unites all of these automatic physical processes (Eishsheim 2015, 112). Vocal phrases are influenced by the length and force of the breath, stimulating a deep awareness of breath and the body among singers and actors.

Parallels between vocal production and somatic practices are evident in practices like yoga, which uses targeted breathing exercises and chanting mantras for accessing intuitive states of being. Breath is often aligned to energy in related somatic traditions. Being able to regulate energy through an awareness of breathing is a key component of expression for vocalists and woodwind instrumentalists. When body actions increase in magnitude and pace, the breath speeds up. Singers learn that breathing inhalations and exhalations can be voluntarily controlled to maximize changes in tempo and intensity (Eishsheim 2015, 112). Then breathing occurs automatically in the background, like an engine powering performance.

In the preparatory phases of her collaborative composition *Body Music*, Eidsheim (2015, 113) notes that bodily composition and shape influence overall vocal production. For example, it is possible to hear a distinct sonic character in the physical gesture of a vocalist's smile as it results in different overtones. Eidsheim's piece, *Body Music* 'makes music by composing actions with detailed attention to the internal, invisible choreography that yields vocal sounds' (Eidsheim 2015, 111). Stance can also influence vocal tone, as an upright posture permits the uninhibited passage of air through the lungs. Yet the singing voice seduces with its beautiful, lilting tones, which shifts listener focus to the resulting sound, not to the individual body producing it, Eidsheim (2015) argues. Her work redirects this focus away from the perfectly crafted singing voice and towards the internal physical architecture influencing the sound. By transferring attention to the actions behind the sound, rather than the sound itself, Eidsheim (2015, 116) emphasizes the potential for singers to release inhibitions and produce sound with greater ease.

Felt sensation, or movement awareness, is vital for vocalists, instrumentalists and performers of gestural systems. Vocal music in particular provides a direct access point for the felt body. Vocalists rely on the inner sensations of their bodies to tune and refine sounds, as well as connect with external rhythms. Diane Hughes (2017, 183) accents the role of kinaesthetic awareness in popular vocal education, supporting singers to convey a desired sound even before listening to their voice. Increasingly, embodied interaction designers are creating works that share similar themes to body-focused educational approaches and pieces like Eidsheim's *Body Music*. Breathing Light (Höök et al. 2016), for example, comprises a pulsing light that directs focus to breathing, highlighting the experience of inhalation and exhalation for users interacting with the design.

Cathy Lane (2020) analyses a series of works by women that feature breath and vocalizations, portraying emotional and physical intimacy through manipulated bodily sounds. These works challenge the conventions defining how women's voices can appear in public spaces. Lane summarizes the tension and symbiosis of the relationship between voice and body: 'The voice is produced and shaped by a body. When heard, it also suggests a body' (Lane 2020, 198). The emotional state and physical condition of a person's body shapes any sounds created, making voice an individually coloured series of tones, rhythms and harmonics that emanate from the unique dimensions and resonating chambers of the neck, lungs and stomach. This link between voice and physiological dimensions is illustrated by the belief that opera singers must be of larger stature to produce a powerful sound, leading commentators to conclude that Maria Callas's mid-career weight loss contributed to a reduction in vocal ability.

When internally sourced sounds are projected publicly with sound reinforcement, they emerge into a new amplified, processed context. The sound goes beyond the body of the performer, affecting listening, self-perception and agency. Kristina Warren (2018) defines contemporary vocal agency as influence and control over the electronically mediated voice. Historically, the female voice is stripped of agency, Warren argues, 'Voice is historically de-agentialised: gendered female, opposed to self-listening and, more recently, made ocularcentric by recording technologies' (2018, 31). Yet in cyborg theory and during the use of technological prostheses like microphones, there is an opportunity for female vocalists to reclaim agency and cyborg status, rather than relying on sound engineers to realize their processing and sound control over a public address (PA) system and through foldback, which enables performers to listen back to their affected voice on external monitors placed on stage. By independently controlling the electronic transmission of their voice, vocalists can enhance their individual sonic impact, moving beyond the existing limitations and separation between vocal and technological labour, Warren (2018) concludes.

The voice, like movement, is transient. Movements give visibility to the hidden, intangible voice. Yet few researchers delve into the relationships between movement and voice in detail with several exceptions like Julia Nafisi (2015) who recognizes the role of movements in articulating and shaping vocal sounds.

Other writers who have emphasized this capacity include Donna Hewitt (2006) and Jane Davidson (2001), who both research how the movements of popular singers influence vocal expression.

Electroacoustic artist and performer, Daphna Naphtali (2017) was encouraged to be still while singing throughout her classical voice training, so as to focus primarily on the instrument. This can become a challenge when learning to use a gestural system that encourages full body movement. For Naphtali it became necessary to overcome the advice to minimise the body and allow physical gestures to help her reconnect with the vibrations she feels when singing, particularly in electronic music where acoustic energy isn't sensed. Pamela Z, whose work explores the intimate relationships between movement and voice, also emerged from a classical vocal background and channels her gestures into constantly reimagining her signature sound.

6

Pamela Z

Pamela Z is a composer, performer and media artist exploring voice, live electronics and video in solo performance. She is regularly commissioned to compose works for chamber ensembles, film and interactive dance projects. Her vocal practice combines the operatic bel canto tradition and extended vocal techniques with movement-based control of sound and video. Z first adopted the BodySynth, a wearable system created by choreographer Chris Van Raalte, which was realized by electrical engineer, Ed Severinghaus in 1994 to manipulate samples and effects. The system is a musical instrument digital interface (MIDI) controller that receives input from muscle activity, leveraging the existing movement language of the performer. Electrodes are attached to the skin on both arms, one shoulder and leg to measure the electrical impulses of the muscles, depicted in Figure 5. As Z explains in *Pink Noises* (Rogers 2010, 210):

> The amount of electricity that's produced by the muscle is measured, and that's sent to a little CPU that turns it into numbers, and those become MIDI data. And that can be interpreted any way that you want to program the computer to interpret it. So it's like having continuous controllers in different muscle groups.

As with Tanaka's work, the system detects muscle impulses that form part of the intention to create sound. This anticipatory type of sensing has led to occasionally unpredictable results for Z in adrenaline-charged performance environments. The mere thought of making a sound could trigger one by sending a charge of electricity through her muscles, initiating intensive streams of data. This capacity to capture the intentions of the performer highlights the link between imagination and sonic outcome in muscle-sensing methods, signalling little separation between the two.

Z has since collaborated with designer Donald Swearingen to develop custom-built systems equipped with ultrasound and light sensors, faders and a footswitch control pod. The devices belong to a suite of N Degrees sensor-based

Figure 5 Pamela Z with BodySynth. Photo credit: Lori Eanes

controllers developed and fabricated by Swearingen, who also wrote the firmware and assisted with programming Max patches to allow Z detailed parametric control over sounds emerging from her devices. The ultrasound controller, called Ute, and the infrared light box, named Mira, can plug directly into any MIDI interface. A more flexible, wearable version, Mimn, pictured in Figure 6, is worn on the hand, equipped with magnetometers, accelerometers and gyroscopic sensors. The portability and ease of use of the N Degrees devices led to the gradual replacement of the BodySynth for processing and looping vocals live on stage.

The foundations for Z's movement-based practice can be traced back to her early musical life, when she became fascinated with delaying and looping

Figure 6 Pamela Z with Mimn. Photo credit: Goran Vejvoda

the voice after purchasing her first hardware delay unit. She embraced the possibilities of instantly creating multilayered performances as a solo vocalist, as she reveals in a recorded online interview in August 2020:

> The thing that really attracted me to doing works with voice and electronics was the fact that I could sing as a single person and that I could instantly create all these layers. So still to this day I would say that delay lines are probably the most essential part of the electronic processing that I do, because so many of the pieces I compose are structured around layers that are created by making loops. Sometimes they're really long loops that don't sound like loops. They sound like textures because they're too long for people to remember them from beginning to end, often times with multiple delay lines at multiple different tempi so that they're rhythmically shifting in and out with each other.

Experimenting with delays and looping offered Z an opportunity to break out of previous creative habits and find a new form of vocal expression. While her first instrument was the voice, the next phase of her practice created a hybrid between voice and live electronics, using a digital delay to achieve increasingly complex vocal textures. The discovery of out-of-phase loops, when the lengths of overlapping loops do not always synchronize and coincide, introduced an intricate layering approach within her performances.

The BodySynth, and later the N Degrees systems, enabled Z to fuse voice and movement more tightly, using gestures to process vocalizations, trigger samples and build compositions in real time. This work built on prior performances with technology where the voice is often heard and filtered through technological means like the telephone. One of her pieces responds to the tortured tone of an early Mac voice assistant, Victoria. Z's 2003 work, *Voci* (Voices) explores the broad possibilities and characters of the voice and how perceptions of individuals are shaped by qualities of the voice. It combines digital effects processing, vocal samples and projected video. The work presents the voice in a variety of contexts, highlighting its role as a marker of identity and emotional health, fluctuating between stylized arias, primal cries and whispers. Technology can either heighten or obscure the character of the voice, removing it from its biological origins. Z delivers duets and choruses of virtual and organic vocal performances with the aid of light and gestural controllers, exploring the various roles of the singing and the speaking voice in communication and character signification.

Embracing gestural systems seemed almost like a natural progression for Z, due to parallels with her vocal practice. Both types of performance do not centre around an external object – they redirect attention to the body itself. Spatial movement and kinaesthetic awareness support professional vocal and gestural performance. This makes the implementation of metaphors in design and performance important for cultivating nuanced bodily knowledge. In 2003, Z composed *Cellovoice*, a piece in which plucking and bowing gestures trigger string and vocal samples with the BodySynth. This metaphoric mapping assists in hybridizing the voice and instrument, incorporating performance gestures that can be easily recognized and interpreted by an audience. It also reflects her instrumental training in viola as a child, learning the nuances of a smooth and consistent vibrato.

The experience of playing the BodySynth allowed Z to refine her gestural language and style over time. She recalls that her early performances with the system were stiff, mechanical and robotic. Practising in her studio, Z learned to control effort through her muscles. Gradually she developed flowing, fluid gestures. As well as refining the quality of her movement by focusing on intention and consciously directing energy, Z developed a unique gestural vocabulary as a method of playing the instrument like activating a pre-recorded sound with the flick of a wrist, explaining:

Sometimes I'm maybe triggering a sample or a note with one hand and then bending the pitch of it with the other, or I might be changing the parameters in the processing of my voice, so I'm singing and I'm using my hands to granulate my voice, bending the pitch of it with the other, or I might be changing the parameters in the process in my voice or do some other processing on my voice, so each piece has a different use of those gestures that are created in response to what's necessary to make those things happen.

Z characterizes most of the gestures she uses as abstract, based on the demands of a particular system or musical work. Playing Mira, for example, calls for vertical hand gestures. Other types of gestures she employs are more visually dramatic and literal, such as the use of typing gestures in the 1995 piece, *Typewriter*, which trigger actual typing sounds. The movement vocabulary for each performance is conceived during the compositional stage rather than improvised on stage. Z works with the voice and processing during composition and rehearsal to arrive at musical ideas and matching movements for the completed work.

The prevalence of gesture in Z's work predates her work with gestural systems. Particular pieces incorporate a sequence of gestures, though they don't trigger specific sounds. In her vocal performance, Z finds that gestures used in vocal production affect tone, even if not directly connected to producing the sound:

When you're a singer, sometimes gestures are really connected to tone production. They're somehow connected to your voice even if they aren't physically producing something, so it was exciting to take that a step further and actually cause these gestures to produce something and learn with the instrument. You gain a certain virtuosity on that instrument that allows gestures to be meaningful as actions that are actually doing something.

The shape and character of Z's gestures are influenced by her operatic training and a love of dance. Merging the classical vocal tradition with multilayered timbres that also reflect on the Afro-American tradition, Z transcends societal expectations (Lewis 2007). Her unique role in hybrid vocal and electronic music questions notions of traditional vocal identity. The unification of voice and gestures let Z combine classical training, extended vocal techniques and electronic experimentalism to expand her identity and challenge cultural assumptions based on established vocal styles and techniques:

When I started working with electronics, I remember a time when I felt really conflicted because I couldn't find a connection between my classical training and other things I was doing. And it was when I started to do more experimental music that I realised I could marry those things and that they could work with each other, and that I could use all the different colours in my palette. I didn't have to relegate this kind of thing to this style, and this other kind of thing to another style. I could make a music that includes all of my tools so that I could use bel canto singing, but I could also do other kinds of sounds – sometimes even in the same breath. I think that working with processing and using the gestural control instruments is super important to my practice. The thing that I would say is most life changing was when I started doing live processing on my voice because that was really freeing in a lot of ways. You want to try different sounds when you can hear them being affected or layered on top of each other, or looped, or granulated, or with a lot of reverb on them. It's like you just want to experiment to find more new sounds. To me at least it seems easier to stretch out.

This sonic exploration liberated Z from a specific style or vocal identity, enabling her to challenge and blur existing vocal archetypes like the virtuosic and powerful operatic performer or boundary-pushing avant-garde experimentalist. Z's most momentous creative discoveries are often prompted by the introduction of a new tool, which can range from a customized gestural system to everyday objects like the popular toy Slinky, which she used for creating percussion instruments in the mid-nineties.

Yet the fusion between voice and technology is not without challenges. Z (2000, 347) lists the incessant recurrence of 'crashes, upgrades and incompatibilities' affecting her productivity with computer-based tools. Technical issues are also an inevitable effect of Z's fascination with mechanical objects that result in different technological configurations for each piece. The theorization of the tool and dominant focus on equipment in live electronic music may suggest a point of division along gender lines. Female artists are expected to use the voice as their main instrument more so than men, Z surmises:

The tool that women seem to be expected to excel in using is the human voice. And when we do excel in that, we do get recognition for it. Cathy Berberian, Diamanda Galas, Joan LaBarbara, Meredith Monk – all these women are very respected and well known for their work with this very technically complex instrument. They are much more celebrated than are any of the men who use extended voice as a main component of their work. But Pauline Oliveros, Laetitia Sonami, Annea Lockwood, Laurie Spiegel, Maryanne Amacher, and the many other women who have done great work in both the designing and using of systems for electronic music are much less likely to be mentioned than their male counterparts. The message seems to be 'If you want recognition for what you do, you need to stick with the tools you are expected to use.' (Z 2000, 356–7)

Z's work with extended voice prompts insights into the gendered aspects of the experimental vocal tradition, questioning the historical convention that ties women's identity and communication style with the body. She often finds that she is usually the only female on an electronic music compilation, and commonly the only artist using voice. According to her own experience as a practising composer and performer, Z recognizes a cultural association of men with the instrumental tradition and women with vocal practice. Is it because men are socialized to feel more comfortable with manipulating tools, Z wonders? The common assumption that the only female musician in a group will be the vocalist is gradually changing, but lingering assumptions with strong historical foundations continue:

> I'm not talking about what's innate or should be but about the way that I think people have been socialised, at least in past generations, where men seem to feel more comfortable improvising and creating really wild or aggressive or exciting noises and sounds if they were armed with some kind of tool or device that would sort of stand between them and the audience, so they weren't just putting their body there, whereas women have always been expected to use their bodies as a way of communicating, as a thing of attracting. So when you're making wild, frantic sounds and they're coming right out of your face, and it's almost like the same thing that men aren't supposed to emote, they're not supposed to cry when they feel sad, but women are expected to wail and moan and all these things.

Another expectation that Z observes is that, to be recognized, female composers need to incorporate the voice somewhere in their work, referring to high-profile artists like Meredith Monk, Joan LaBarbara, Laurie Spiegel, Laetitia Sonami and Laurie Anderson. Without a vocal component in their compositions, many female composers have been ignored in musicological anthologies and histories, Z notes. On the opposite side of the spectrum, she can list few works featuring extended voice for men, with notable exceptions like David Moss, Roy Hart, Dimitrio Stratos and Jaap Blonk. Yet she can easily rattle off a list of female artists such as Joan La Barbara, Cathy Berberian and Diamanda Galás. Z theorizes that artists are rewarded for following the norm. Yet the Wikipedia entry on extended voice lists predominantly male composers and performers including luminaries

Karlheinz Stockhausen to Trevor Wishart, indicating under-representation of women's contributions in written accounts of the extended voice.

In Z's experience, it is as if men in the avant-garde are more comfortable with experimenting sonically with a physical instrument, which acts as a form of armour. It is considered more acceptable for women to directly use their bodies to create new, unexpected and sometimes uncomfortable sounds than men, Z observes: 'When an artist uses his or her own body as an instrument, it is like being naked.' If a musician arrives on stage exposed and vulnerable, with just their voice and body as instruments, it appears that audiences are more accepting of female performers in this position, according to Z. Another general observation she makes is that women may be less confident around technology. Yet this particular field offers an opportunity to disrupt any remaining assumptions, Z argues, as 'the type of electroacoustic music that combines vocal practice with electronics might be viewed as a way of exerting both feminine and masculine qualities in the performer' (Z 2000, 360). Both voice and movement can be interpreted as traditional female domains because they foreground the body (Bosma 2013). Gestural music offers a seamless connection with composition, a traditionally male domain. The voice and electronic devices are integrated and not separate like the singer–instrument accompaniment model. There is a fusion between instrument and body, unlike in any other type of performance practice. Z is against separating live electronic hardware and the physical body, asserting that 'whether the instrument is acoustic, electronic, analog, digital, flesh and blood, or some combination, a tool is a tool is a tool' (Z 2000, 361).

Z's busy touring schedule and ongoing commissions indicate a composer and performer in demand, balancing abstract ideas, unique vocal expression and customized technological set-ups for solo and ensemble works. Her dual identity as a composer/performer places her in between two activities often divided by gender, as Hanna Bosma (2013, 217) argues in relation to electroacoustic music:

> Vocalist-composers use recording, amplifying and sound processing for the manipulation of their voice sounds, for structuring and composing, for performance, and for fixing the composition and disseminating it. They are seen on stage mastering technological equipment. ... Thus, they combine feminine cultural practices of singing and performance with the masculine cultural domains of avantgarde, authorship, composing, and technology.

When Z first started performing as a soloist with technology, audience members would question who set up and designed her systems; however, now they were more likely to ask if she designed and built them herself, indicating a gradual change in attitudes towards women performing with technology.

Working with gestural systems has opened Z up to merging styles and art forms. Not confined to a particular genre, Z unites the electronic and acoustic parts of her practice, combining bel canto singing with extended vocal techniques and video projections, thus expanding the colours in her creative palette. Visuals are a key component of Z's works, like *Sixteen Actions* (2013), in which she captures sampled video fragments of her face with twisting wrist gestures, then manipulates and rearranges them, scrubbing through the captured samples with sliding actions. As one of the high-profile users of real-time video manipulation software, Isadora (Coniglio n.d.), Z extends her Max patches to incorporate visual feedback:

> In my large-scale solo performance works they always have some degree of projections involved, some of which are interactive and some are just things that roll in. And then in my solo concerts I often also use projections and I do have a couple of pieces that I do where what's happening on the screen is controlled by gesture.

Video can also be found in her commissioned chamber works, such as a piece for Del Sol Quartet, called *Attention* (2016). The sonic qualities of the string quartet piece are reinforced by fixed media in the form of multichannel audio and visual imagery:

> I composed this entire video that is absolutely required in order to perform the piece, and for the first section of the piece, what's on the video is like a graphic score. The quartet actually turns away from their music stands and face the projection and they play according to what they see on the screen. That particular part I think for me, the first idea I had was this image of these drips of water dripping down and thinking about that connected to a glissando, and so I actually made the image first and it became a structure for the piece.

The visuals amplify the tonal quality and movement of the sound through matching natural and symbolic images that give form to Z's commentary about the constant disruptions in focus caused by endless chatter and mobile phone notifications that characterize modern life, underscoring the pronounced visual presence of both Z's solo performances and commissioned compositions.

In terms of how Z's practice has shaped her approaches to movement, the BodySynth became like a biofeedback device, offering information about the body's underlying functions and how to control them. The system, in monitoring the electrical activity that causes muscle contraction, reflected back to her how much energy she was exerting through her muscles and how to direct that effort more purposefully:

> All of these gestural instruments are like biofeedback devices because when you do something, you're generating numbers and so you get some kind of instant result from what your motion is doing, so the BodySynth was using the effort from the muscle.

As previously mentioned, the effect of the rise in adrenaline during performances generated much higher numbers in the control data. As a result of this experience, Z learned to become more still and centred during her performances. This engagement with the system affected her overall awareness of the feelings governing her body:

> The quality of my movement changed radically, and it was because of learning how to be, and how to have intention. ... It was like learning to play any instrument. It was just learning to have control over the tension in your body. Because that instrument was so sensitive, if I turned the sensitivity all the way up, I could make a sound just by thinking about moving my arm, because it's going on your electrical impulses and when you think about moving your arm, your brain sends an electrical impulse to that muscle to prepare it to move.

As a result of long-term engagement with the BodySynth, Z's gestures became more refined and deliberate, resembling communicative gestures that accompany speech and intricate string instrumentalist actions, like bowing and plucking invisible strings:

> Using that instrument, suddenly more studied and much cleaner, isolated, gentle and slow qualities came into my gesture – suddenly it was like liquid. That was not a matter of me looking at myself and saying I wish my movements were more fluid. It was me trying to learn to play this instrument, and through that, I wasn't even realizing it. Then I would have people come up to me after the show and go oh, your hand movements, your gestures are so beautiful, did you study Indian dance and I said no, I just learned to play this instrument. Then later I started comparing videos of myself using these kinds of instruments to videos of myself before I had used these instruments. The difference between the quality of my movement was extreme.

This transformation complemented her other somatic experiences, including training in modern dance, classical ballet and the Japanese dance tradition of

Butoh, under the guidance of Butoh master, Kazuo Ohno in his Yokohama studio. This latter experience was channelled into the large-scale work, *Gaijin* (2001), where Z performed with three Butoh dancers, expressing feelings about being a foreigner during her six-month artistic residency in Japan. Embracing the slow movement style of Butoh, Z developed a deeper appreciation of the temporal qualities of movement and learned to detect subtle changes in her body while it explored space in slow motion. The physical understanding that arose from those performances and formal movement training is akin to the body awareness expert vocalists develop. Vocalists *feel* a pitch in their bodies rather than relying on an external object for physical feedback, making vocal practice a close relative of gestural practice, Z reflects:

> Their body is their instrument, and so already I'm using my body to make sound, and I'm having to relax or stress cords to get them to do different pitches. I'm having to use my breath in a certain way – all of those things. I think wind players have a certain amount of that, and actually anybody who plays an acoustic instrument has involved their body in some way, but it's usually involving your body in relationship to an object, whereas singing is involving your body not in relationship to another object and so then gesture controlled instruments I think are the closest to that because you're controlling something with your gestures, but you're not touching something or holding something.

Z works with the body in terms of movement and vocal expression, observing how the resonating chambers of the face, including the nasal cavities, contribute to the projection of an individual voice – a phenomenon called the mask. The technique involves sympathetic vibrations of the bones in the face, composed of the three main vocal resonators – nasal, pharyngeal and oral – to achieve a bright and powerful tone. This approach recognizes the instrumental capacity of the body, capitalizing on physiology to enhance the impact of individual vocalizations. Z's advanced vocal knowledge offers a nuanced view of how the voice serves expressive and theatrical purposes. She matches this with a willingness to experiment with movement and a range of sensors to expand her vocal identity and compositional palette.

7

Julie Wilson-Bokowiec

Julie Wilson-Bokowiec, whose work spans theatre, digital performance and interactive media art, performs with the Bodycoder, designed collaboratively with Mark Bokowiec in 1995. The wearable full-body gestural controller is used in a range of applications, including vocal performance (Bokowiec 2011). The flexible system can be adapted for varied contexts and individual pieces, rather than acting as a fixed instrument. Wilson-Bokowiec characterizes her work with the gestural system as both musical and theatrical, creating hybrid interdisciplinary events that place audience experience foremost. Her exposure to influential interdisciplinary artists like British dancer, actor and choreographer Lindsay Kemp and non-Western movement disciplines including Butoh has contributed to this broad artistic focus. Professionally trained in contemporary dance, Wilson-Bokowiec also received vocal tuition from a young age, driven by early aspirations to become an opera singer. She draws on dance training in classical ballet and the modern dance tradition informed by Merce Cunningham and Martha Graham in her early pieces combining dance and technology, sculpting her physique to execute technically demanding choreography. Later works went on to feature voice, freeing Wilson-Bokowiec from a singular identity as a dancer.

Wilson-Bokowiec entered the field with the 1993 work *Navigator* for dancer and multiple sensors for controlling sound and lighting events. Mark Bokowiec was commissioned as a composer for the full-length theatrical work, which required a complex hardware set-up incorporating Rank Strand's novel musical instrument digital interface (MIDI) lighting desks and the installation of a multitude of wires in a large-scale theatre. The piece aimed to liberate the solo performer to navigate several interactive image and sound technologies, triggering multimedia cues without external technical interventions. The complexity of the set-up and restrictions on the dancer to move to particular spots on stage highlighted a need for Wilson-Bokowiec to pursue greater freedom in future works, as she observes during a recorded Zoom conversation in May 2020:

> Liberation on stage with technology has always been a thing that we really struggled with for a long time. It's been a motivation. It is a relationship and a theatrical problem to do with how do you get to the point where you are fully in dialogue without encumberment with technology? That's the problem people encounter. They make a really interesting piece but then they don't persist. They let things go. Or they find the problem insurmountable so they don't continue.

Adopting the lessons of the first piece, the next generation of system development embraced a wearable, wireless approach by attaching all the technology Wilson-Bokowiec required on stage to the body so that she could move freely. The Bodycoder was created in 1995, prompted by her desire to make the system invisible and transparent, opposing the 'predominantly demonstrative use of technology' in live electronic music:

> I choose to hide the technology to a large extent underneath a costume, because I'm not demonstrating the technology, I'm actually creating a piece of work. I don't want the sight of technology to get in the way of enjoying a piece of performance. … For some performers, that feels quite negative, to actually not have the technology visible because the technology's somehow validating what you're doing. Obviously if I showed the technology and it was self-evident that everything was live, I would potentially get over those ideas that stuff's being recorded. But at some point, the audience has to make that leap with you.

Wilson-Bokowiec sees a gendered aspect to techno-fetishism in live electronic music, where the glorification of hardware and display of sophisticated technical configurations takes priority. Her energy is not directed towards a particular object, echoing Z's approach that places the body at the forefront of pieces rather than technology.

Following a string of successful performances throughout Europe supported by the British Council, Wilson-Bokowiec's work was increasingly viewed through a dance-and-technology lens. Critics and audiences recognized her primarily as a dancer, some even questioning whether she was a musician. Wilson-Bokowiec made a conscious decision to transform her performative identity by transitioning from predominantly dance to vocal works. In a set of interactive vocal works composed in 2007, *Vox Circuit Trilogy*, she decided to stay rooted in the same position, minimizing her physical skills to escape from being viewed through the narrow prism of a dancer using technology. For the second piece in the series, *Suicided Voice*, Wilson-Bokowiec sought to expand her performance presence by constraining the scale of her movements and inventing new ways of moving beyond ballet and contemporary dance choreography:

> I was going to see what happens when I take the technology away from my legs, and *Suicided Voice* was the resulting piece. And I hadn't used my voice for a long time up until that point. So the other thing that was great is that I wasn't a lieder or an opera singer. I'd forgotten much of that training. But my body hadn't forgotten the dance training. Yet my body was so well trained, it could only put itself in the shapes of contemporary dance or ballet. So I thought, by preventing it from doing that, and from actually not dancing, maybe my body could forget that after a while and come back to the body later and see what would happen. Hopefully it wouldn't remember.

This conscious identity shift led Wilson-Bokowiec to a completely new direction of rediscovering and highlighting her voice, focusing all the sensing technology on her upper body rather than her legs. Out of this process she found a fresh way of working and developed a novel physical language that was not tied to a particular art form. Embracing her vocal practice again after a ten-year hiatus allowed Wilson-Bokowiec to portray a more multifaceted relationship with technology suited to the emerging nature of interactive digital performance, asserting: 'You have to break the genres or move away from the genres so that the audience can see something different. So that was the reason for the vocal work, really.' The piece symbolizes the power of the transformation, 'suiciding' the acoustic voice to digital signal processing within Max/MSP, defying gender-specific pitch registers, amplifying the resonances of rich extended vocal textures and manipulating computer graphics in real-time.

In preparation for a piece, Wilson-Bokowiec and Mark Bokowiec, who contributes to composition, electronics and software design, combine collaborative design sessions with rehearsals. All practices are recorded 'to find a physical vocabulary simultaneously with the sound vocabulary', Wilson-Bokowiec explains. Figure 7 depicts a rehearsal set-up for the piece, *Etch*, the third piece in the *Vox Circuit Trilogy*. Wilson-Bokowiec wears the Bodycoder on-body interface with gloves, bend sensors on both arms, a neck sensor, head cam and microphone. Live sound processing within Max/MSP on the foreground computer is also sending live MIDI information to Xpose VJ software by ArKaos, running on the computer in the background for image manipulation. Multiple sensors are activated by the fingers of the right hand, while the left hand

Figure 7 Rehearsal for *Etch* © Julie Wilson-Bokowiec

controls live sampling and navigates the software, either mapping particular gestures to single effects like filtering video colours, or more complex mapping of gestures to multiple processes that incorporate recording of audio, frequency filtering and pitch control. Granular synthesis, a process in which short samples, or grains, of sound that are segmented, reordered and manipulated, is used to create an array of sonic textures, which are sometimes reorganized randomly. Wilson-Bokowiec generates all sound live, requiring extensive improvisation and prototyping in preparation for each performance to determine the sample size and mapping of particular processes to specific gestures:

> We spend an awful lot of time talking initially, because once we start working with the system, I'm in the system all the time. So studio work is like days, months, weeks inside of the technology. I'm inside of the interface trying things and tweaking things. What we don't do is, we're going to do this, Mark goes off and does a lot of programming, I compose a lot of vocal music, and then we come back and put it all together. What we tend to do is be in a space for a very long time with all the technology and we tweak it and bend it, and a piece emerges out of that. We might find a way of working that we particularly

like and I'll improvise physically and vocally with that for days. And we record everything.

Wilson-Bokowiec reflects on video recordings of these exploratory sessions, using her well-honed kinaesthetic memory to recall specific positions and movement-sound associations refined during rehearsals. This physical immersion lets her evaluate the strength of links discovered between movements and sounds:

> What I'm doing when I'm going through this intensive process is I'm trying to find a physicality that embodies that or feels right to express that kind of sound. There's a language there. Each piece has a very particular language, which I'm trying to find the logic for, and I can't tell you what the logic is. It's a feeling really. Also, I'm extremely aware, in working in that way, that I'm trying to forward imagine what that might look like on a stage space, so I can sense the quality of the sound and physicality, and I'm trying to put myself in this imagined character in this space.

This feeling-based approach echoes the insights of artists, Tanaka and Z, whose engagement with intimate muscle-sensing systems contribute to a heightened appreciation of inner physical sensations arising in design and performance. Although this felt feedback is difficult to describe and pinpoint, it appears to be a major form of guidance for musicians engaging with gestural systems. The need to combine the related activities of composition, software development and defining a gestural language are critical to supporting this exploration, allowing inner feeling to manifest in each aspect of Wilson-Bokowiec's practice. She believes in combining composition, programming and physical experimentation so that all three elements evolve in tandem, rather than commencing with fixed design ideas or code:

> I think it's important for us that all those decisions take place simultaneously. Then you're not favouring one thing over another and the truth and logic of the language that you're trying to articulate comes directly out of the making process, which is neither a musical composition or a dance composition or a vocal score, but a combination of all three.

Improvisation with the system is a key aspect of refining and adapting the mappings between sound and physical movements for each piece. Kinaesonics, a term that blends kinaesthetic understanding of movement with sonic meanings, captures the role of performer experience in defining sound-movement mapping associations (Wilson-Bokowiec & Bokowiec 2006).

In terms of performance technique, Wilson-Bokowiec displays similar rigour to Z, precisely and consistently re-enacting selected movements through extensive practice: 'I pride myself on strong technique. Strong technique within the system is being able to hit everything, to be able to perform an intentional piece, rather than improvising.' Wilson-Bokowiec processes her vocal with limb gestures to produce a unique hybrid texture of acoustic and digital voice. Exploring extended and operatic vocal techniques, her range has expanded over time, gradually moving beyond a high soprano range to a lower register. Wilson-Bokowiec looks specifically at ethnic modes of vocalization in order to create different timbres. Intricate overtones, multiphonic textures emerging from the throat and broken vocal sounds are captured by a high-quality headset microphone. She aims to enter into the sound, examining and manipulating it by transforming tiny grains to create markedly different textures from the source material of the acoustic voice:

> When you're then imposing digital processing on that, you're getting layers on layers of timbre, which is so interesting. It's incredible. Then if you actually listen to those layers, if you've got an ability to go inside those layers gesturally so you can pinpoint little grains of sound within that, it's spectacularly intimate.

This timbral exploration exposes sounds that are often hidden, tucked into the back of the throat. Once amplified, processed and comprehensively explored, these sounds are exposed to detailed listening, notes Wilson-Bokowiec:

> What we're interested in in terms of the aesthetic of our work is that internal nature of sound, so we're not interested in playing tunes, so to speak, but in being inside the sound, in a sort of very deep listening way. The feeling for me is about vocalising and exciting certain sound processes and when they come back in the system, it's about being able to move inside of that sound gesturally. It's actually like going inside of your voice and moving inside of the space of that particular sound ... when you discover something and it sings, it really makes my hair stand on end – the vocalisation within a certain kind of sound process – and suddenly it's an amazing thing, and that thing is dimensional, in terms of, you can move it within a space gesturally, you can move aspects of it gesturally. I can have a relationship with that sound physically within the space.

This approach is evident in the work of Eidsheim (2015) discussed earlier, which is focused on exposing the inner workings of the voice, and bassoonist Joanne Cannon from the Bent Leather Band, who amplifies the subtle inner breath sounds of the bassoon that are often lost in ensemble performance.

In performance, Wilson-Bokowiec does not believe in making the system too easy to control. To bring out the intensity of a performance piece, like the layered interactive work *Voc't (Ritual)*, composed by Mark Bokowiec (2011), Wilson-Bokowiec manipulates predefined compositional structures with a sensor array of bend sensors on the torso, arms and legs and a navigational glove. Sweeping gestures at specific angles apply granularization, looping and filtering to the voice (Bokowiec 2011, 41). The sound is distributed through an eight-channel speaker array surrounding the performer, whose gestures activate various spatialization routines or automated changes. During rehearsal, mapping processes are adjusted to introduce complexity into the system controls that require the intensification of motor and listening skills:

> The hearing to movement to sound is so intense, that I'm really straining to hear, first of all, that little thing that I'm controlling. But also, invariably I'll put that on the smallest lever, which as you know because you work with gestural things, the smaller the lever, the more difficult it is to control it. So the more intense the movement, the more focused your body is, the harder you work on controlling something. So you might be on a massive lever but you're actually working with a centimetre of movement to do something quite big in a soundscape.

Alternatively, Wilson-Bokowiec attempted to pinpoint a particular timbre in a lengthy sample she'd recorded live rather than trimming the sample length and vocalising that particular section to prevent the performance from becoming too effortless and automatic:

> *Voc't (Ritual)* is a really good example of that, because it's an extremely hard piece to control, and there are so many layers that I'm gesturally controlling. It's a really, really intense performance experience for me and sometimes I would make things, we would design things in the system that were particularly hard to control in order that I had to really dig deep in performance and you see those moments cause I suddenly go urgh! It's a battlefield.

The conflict metaphors that Wilson-Bokowiec evokes to describe challenging performances like 'battlefield', 'combat mode' and 'going to war' mirror Sonami's thoughts on balancing between creating a dialogue with the system and wrestling control from the machine. Wilson-Bokowiec capitalizes on this tension to sharpen her focus, finding difficult performances ironically rewarding. The intense concentration they require translates into a highly refined gestural language. Precise finger and wrist movements indicate learned gestures that can be consistently reproduced to invoke fine-grained changes to the qualities of the voice.

Also affecting the degree of control is the setting in which the work is performed. Many venues hosting bands, soloists and ensembles are unaccustomed to the specific demands of digital performance. Wilson-Bokowiec views the multichannel speaker or diffusion system as a fundamental part of the Bodycoder system. It provides detailed auditory feedback required for pitch and processing control. When a different type of speaker and foldback system is supplied or the acoustics of a particular venue alters the overall sound substantially, Wilson-Bokowiec draws on body memory refined through rehearsal to recall feelings linked to certain pitch positions and types of vocal expression within the body. Applying highly attuned bodily awareness and control to producing gestures, tones, timbres and pitches at will, she compensates for the absence of reliable and detailed auditory feedback:

> One of the reasons why we rehearse a piece so much is because it's to do with body memory, particularly body memory around the voice production. I've done works where I've not had any feedback, so I can't hear what the audience is hearing. The only thing that I have then is the kinaesthetic memory of my body and my voice production.

Kinaesthetic awareness is a skill many vocalists performing in amplified spaces must develop, as sound reinforcement and foldback systems can vary wildly between venues and events, leading to auditory-motor disalignment where the link between actions behind the sounds is disrupted (De Souza 2017, 85). This gap between what the body is doing and hearing is heightened for the vocalist when the voice is processed and the unaffected, acoustic signal is not present or obscured within the mix.

Despite these challenges, the Bodycoder system offers Wilson-Bokowiec a way of transcending and moving into new, unexpected places with an audience: 'For me the act of performance is a really fundamentally sacred event, and I expect to go places in front of an audience when I do that, and the system facilitates that.' This act of transcendence is aided by solitary reflection after each performance, processing and internalizing the feelings that arise throughout live work, thus reaching new states of understanding. This results in a novel appreciation of the voice, making it prone to reinvention as it is explored through gestural manipulation once leaving the body.

Despite the shift to vocal works, Wilson-Bokowiec continually draws on her dance background in gestural performance: 'My initial training was really invaluable to me. I'm totally aware of my body in space at all times,' she observes. As part of her personal movement practice, Wilson-Bokowiec gardens and attends modern dance classes in Martha Graham technique and Iyengar yoga to move through an intensive pattern of physical forms under the guidance of a skilled teacher, while assisting with core strength. She considers these activities as separate to priming her body for performances with the Bodycoder, which consists primarily of rehearsals over a specific number of weeks where she can inhabit the system fully.

One of the main problems of gestural performance being an emerging art form is that audiences don't always understand what's occurring on stage and make inaccurate presumptions, Wilson-Bokowiec reflects. She is well suited to this murky realm, unafraid to embrace the unknown publicly:

> I like things that are really hard. I like a real challenge, and I'm not afraid of putting myself right on the edge of danger in front of an audience. I kind of like that. The other thing that really interests me a lot about seeing myself in performance is that that intensity of where I'm at in front of an audience, really, really shows on stage. It's an intensity that's extraordinary I think in the performance. I like showing the audience that. The audience can see that I'm completely inside of something.

Feeling grounds this perspective, framed by heightened awareness during stage performance and post-performance reflection:

> I'm extremely aware of my thought processes and what my body's doing and how I'm feeling in performance. I'm really memorising every sensual thing that I'm feeling in performance, and when I come off stage, I need to think about that for a long time before I speak or see anybody after I've done a show, cause I need to internalise what happened and how I feel about that, and also what just happened with my relationship with technology, with the space, and with my hearing of my voice coming back, or the voices in the space, because some of the voices that I create with extended techniques, once they've been through the processing, are in the space or they're entities. They have their own particular characters.

Wilson-Bokowiec's 2018 performance of *Pythia: Delphine21*, composed by Mark Bokowiec, merges the mythological and embodied, evoking the ancient Delphic oracle through which the god Apollo speaks. A visualization of the technological, cultural and site-specific stratigraphy of the work is displayed in Figure 8. The schematic is a reference map and graphic manifesto rather than a score. The voice

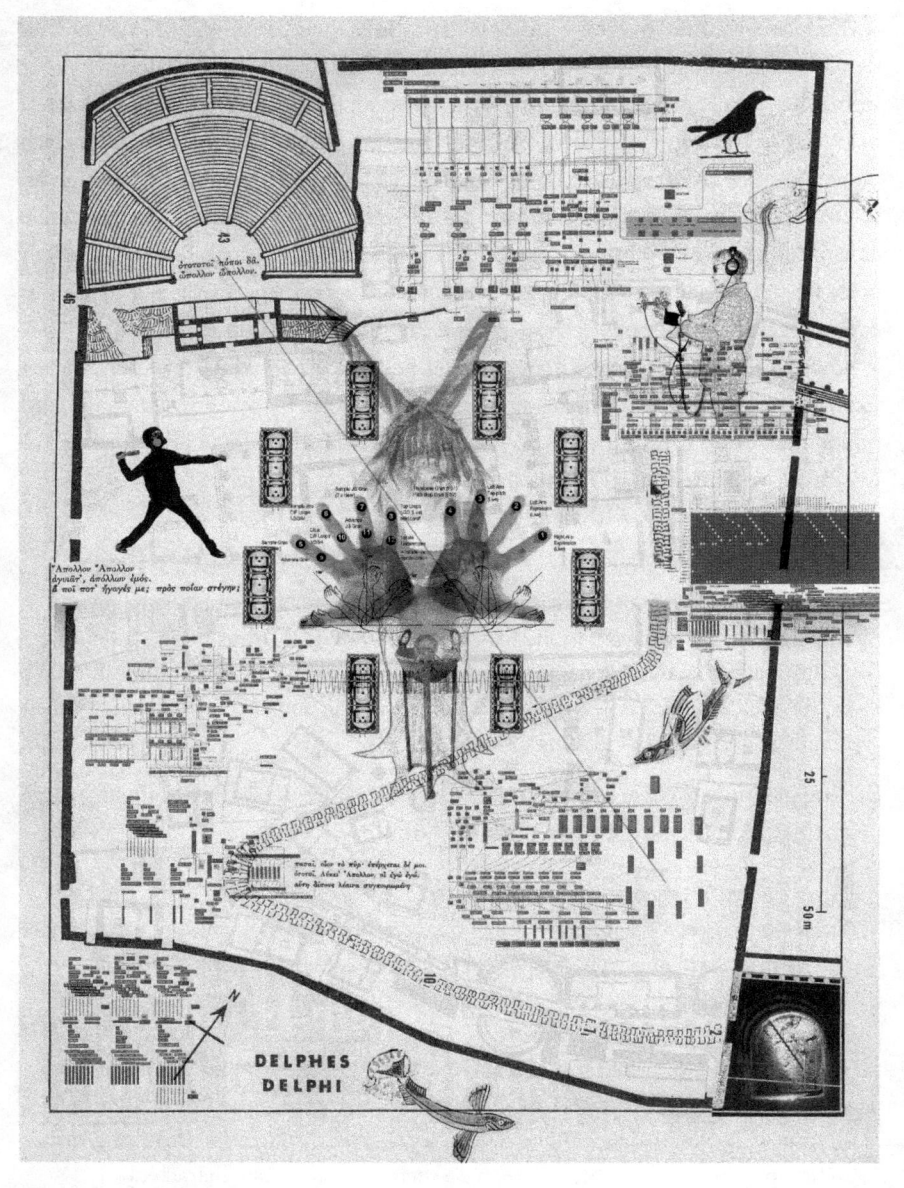

Figure 8 Performance schematic for *Pythia: Delphine21* © Julie Wilson-Bokowiec

is transported within the piece, and new identities are realized as the traditional tonal markers of gender and body patterns are reimagined. Wilson-Bokowiec expands beyond gender-specific vocal registers using extended vocal techniques and ethnic modes of vocalization to reinvent the voice. The piece accentuates the fluidity of Wilson-Bokowiec's work led by her desire to inhabit the body and also launch beyond it to transcendental and mythological states. She transmits mysterious oscillations and drones as a sole figure clad in a handcrafted drapey costume of wool and borrowed objects, hiding her physical form and identity. Cicada and bird sounds filter and soar in the real-time composition. The voice rises to meet the natural textures, layering piercingly high and low rumbling utterances. Her body is never fixed, founded on the discovery of new musical forms and sonic identities while expanding on the rituals of performance with the Bodycoder system.

8

Lauren Sarah Hayes

Lauren Sarah Hayes is a Scottish musician, sound artist and researcher who improvises with an assemblage of hardware and software adapted to her physiology and prior instrumental experience. The continually evolving hybrid digital/analogue live electronic set-up combines commercial analogue synthesizers, musical instrument digital interface (MIDI) controllers, drum machines and vocal processors. Handheld gaming controllers incorporate tactile and force feedback, allowing Hayes to leverage her skills as a classically trained pianist, accustomed to experiencing tangibility and pressure sensitivity through her fingers. Hayes develops customized software for individual works using Max/MSP and Pure Data (Puckette n.d.), incorporating her history 'not only as a musician but as a human being, since I started hearing sound and playing the piano at four years old, and all those different behaviours and skills that were learned and have led to this', Hayes reflects in a Zoom conversation recorded in June 2020.

Transitioning from piano to live electronic performance informed Hayes's goal to highlight physicality in her works. Implementing the vibrations and rumbles of haptic feedback in her set-up, Hayes aims to realize music that reflects 'effort, or actually being able to feel electronic sound, which doesn't have any resonating strings or tubes like an acoustic instrument'. Gloves or game controllers like joysticks let her physically interact with the electronic sound through vibrotactile feedback:

> The glove was actually a haptic sound that allowed me to feel aspects of the electronic sound. It's not actually a controller, and most of the stuff with haptics has been vibrotactile feedback, but I also worked with the Novint Falcon which has force feedback as well. All this is from PC games and the haptic stuff that I use. Really I work with this PS2 controller. In 2007 I went and got this generic five-pound controller and then it just grew from that. I like things that are portable, that I can travel with. If it breaks I can get another one.

Hayes's interest in the interchange between sonic and tactile experience led to experimentation with low-cost controllers with inbuilt force-feedback functions to offer haptic resistance associated with traditional musical instruments. For Hayes, there was a limited sense of direct connection when gazing at a computer screen. Physical engagement with a system guided by touch was far more likely to yield ideas: 'It's about the playing and the exploration and building systems that I can feel and navigate.' In early works, haptic feedback allowed Hayes to ignore the screen completely, instead receiving direct skin feedback when a change in the score was due. She later extended this work in a related research project, contributing to the development of haptic feedback to improve musical perception and hearing experiences amongst wearers of cochlear implants through sensor-aided pitch cues (Luo & Hayes 2019).

Hayes's training as a pianist informs haptic system design that reintroduces this physical feedback into digital musical instruments (DMIs), focusing specifically on the hands and fingers. Her extensive exploration of touch-based gestural systems dates back to her PhD thesis, 'Audio-Haptic Relationships as Compositional and Performance Strategies' (Hayes 2014). Her experimentation with vibrotactile feedback draws out the vibrational basis of sound. For Hayes (2019a, 55), her work is a direct response to the virtuality of constructed systems in which 'we are no longer dealing with the physical vibrations of strings, tubes, and solid bodies as the sound source. Rather, our material is the impalpable numerical streams of digital signal processing and control data.'

Hayes also works with prepared piano and electronics in performance. Her earlier works involve digitally augmented pianos in which tactile engagement is extended and heightened, such as *kontroll* (2010) for prepared piano, live electronics and self-playing snare reacting to the rhythmic elements of the piece with preprogrammed patterns. A vibrotactile glove sending information about various sound parameters and score or clock position to one hand haptically allowed Hayes to focus solely on magnifying the harmonic and percussive qualities of the acoustic instrument without needing to refer to a laptop screen. In recent live streaming performances, Hayes also performed several pieces for piano and live electronics, as illustrated in Figure 9, which depicts a video still from a live streamed off-site performance for London venue IKLECTIK in 2020.

Sometimes I work with a Gametrak or MIDI controller and piano and electronics. There are two diverging modes I can go in. It can vary between doing more ensemble playing with a freer software system, whereas with a solo thing I might

Figure 9 Lauren Sarah Hayes – prepared piano performance video still

want a clearer structure. I recently got back into that because I think for the live streams I had a piano here so that's what I chose to do. And there's an improv patch, microphones and controllers.

As part of her movement explorations, Hayes (2019a) discovered the utility of micro-gestures in producing significant changes within live electronic music. Unlike piano performance, where effort matches the intensity of sonic events, the minutest gesture in electronic performance can activate large-scale sonic outcomes. For Hayes (2019a, 58), tactile and haptic controllers bridge the sense of disconnection between minimal actions and momentous sounds amplified through loudspeakers. They deliver a sense of immediacy to her hands, allowing her to feel more intimately connected with the sounds she produces. Hayes shares this interest in the relationship between touch and sound perception in her teaching. In a 2019 workshop for the Yorkshire Sound Women Network (YSWN), Music for Ears and Bodies, Hayes offered techniques for experiencing sound through feel with bespoke technology. She encouraged participants to explore how feeling their music might provide an alternative insight into their music or compositional process. Hayes also asked them to consider how this tactile approach could offer new potential strategies for building DMIs.

Listening is another activity that Hayes accesses to appreciate how sound can be felt and appreciated in an embodied way. Hearing and listening are processes that are felt within the body, vibrating the inner bones of the ear in a tactile fashion. She draws from the observations by virtuosic deaf musician, Evelyn Glennie (Glennie, Gilman & Kim 2018) in her 2019 article, 'PARIESA: Practice and Research in Enactive Sonic Art', noting that under 20 Hz, sounds are no longer heard but felt. The rumbling bass inside a nightclub can be most strongly felt in the lower body and stomach as inescapable, invasive vibrations. Glennie's finely tuned sensory awareness demonstrates that hearing is a way of feeling sound, which vibrates not only through the structures of the ears but also through the organs and bones. Headphone listening challenges this association, not only turning listening to music into a private experience but also preventing listeners from hearing sound directly through the whole body. It converts listeners' connection with sound into a far more cerebral and distant experience.

When designing systems for her own use or for individuals with sensory impairment or learning difficulties, Hayes carefully considers users' physiology and history, searching for 'meaningful points of friction or resistance as places where the potential for expressive musical engagement lies' (2019a, 58). Hayes dispenses with the trend towards effortlessness in DMIs, instead developing performance systems that emphasize a deeper physical connection to sound. She designs systems that prompt exploration and play, adopting an open-ended approach. In her work as a performer, educator and in designing systems for individuals with learning difficulties, the following factors feature highly: 'improvisation, exploratory play, and do-it-yourself instrument building are key strategies for engendering creative musical activity' (Hayes 2019b, 449).

Hayes (2019a, 54) also incorporates metaphors and imagined agencies in her systems. The unpredictable system she uses creates a dialogue between herself and a hybrid machine set-up. Hayes desires a system that responds with unexpected results at times to avoid imposing limits on creativity. Designing a set-up that she feels compelled to develop with, 'work with and navigate' is another key priority:

> It's so much about my physiology, the way I like to move, or my aesthetic choices, or thinking about how physical boundaries might meet with digital boundaries. It's so personalised. It's enactive in the sense that there are so many parameters now that I can't represent it in my mind. It's not a representational system, I don't think. When I'm playing it and changing a number of pulses and this envelope and that envelope, it's a system that I can navigate through.

This approach stems from enactive music cognition, which is informed by highly influential enactive theory that emphasizes the central role actions play in shaping perception and conscious thought, as well as sensory and motor processes (Varela, Thompson & Rosch 1992). Hayes positions her performance work within this field, viewing it as an evolving, embodied process influenced by interdisciplinary collaborations and piano training, branching into neighbouring areas of health and pedagogy. The dynamic interaction between hardware, software, audience, environment and loudspeakers creates an ecosystem where simple mappings acquire a new complexity. Hayes stresses the need for a system delivering 'things that are thrown up for me to respond to them through improvisation, maintaining that kind of struggle and tension, where something to play with and against is the unpredictability for me'.

In her live electronic performances, Hayes flits between devices, focusing her whole body on a handheld game controller, while embodying the contours of particular sounds with subtle sways and shoulder hunches. She assembles intricate atmospheres of electronic noise, sometimes screeching, distorted, rumbling and apocalyptic, as in her 2018 solo performance at London creative space IKLECTIK. Completely improvised, Hayes's work evokes experimental noise music textures and remnants of pop and techno. Hayes (2019b) offers an extensive historical overview of improvisation to frame her work, analysing a tendency to highlight instrumental mastery and virtuosity in the successful enactment of the improvisation process. Hayes warns that this approach can simplify and prevent a deeper, practice-based understanding of improvisation, explaining, 'I was just getting annoyed at all the cognitive scientists and philosophers who talk about improvisation as this thing where we get to this point and reach a state of flow, and everyone's on the same wavelength, and I thought, you've clearly never improvised. It's not what happens'. Contrary to prevailing theories that a musician needs to achieve a virtuosic skill level to enter a state of unconscious flow, Hayes views improvisation, like broader live electronic music, as a practice that need not be confined to expert/novice dichotomies. She instead embraces the enactive view espoused by Varela, Thompson and Rosch (1992) to capture the 'flexible and diachronically emergent capacities of humans' representing a range of 'holistically embodied sensitivities' (Hayes 2019b, 451–2). Enactive philosophy takes a wider range of embodied performer experiences into account, opening up the vital practice of improvisation to musicians of differing abilities in collaborative and interdisciplinary settings.

The 2018 Laboratory for Laptop and Electronic Audio Performance Practice (LLEAPP) for early career researchers and postgraduate students provides a foundation for testing these concepts. Hayes employs Pauline Oliveros's (2005) deep listening exercises to prime participants for interdisciplinary workshop activities and promote sensory awareness and critical sense-making through available technologies. Students with no experience of improvisation are brought together to perform pieces and develop their sensorimotor skills, exploring relationships between individuals, instruments, technology and the space itself. In another context, students participate in a laptop orchestra, contributing mappings from separate laptops to form an overarching collaborative instrument.

For Hayes (2019b), embodied participation in improvised musical activity has attached cultural connotations. Her pedagogical approach involves drawing on the socio-historical influences of the improvisers. Collaborative interdisciplinary performances aim to merge the various influences of numerous performers, offering an emergent historical perspective to complement skill-related theories of improvisation. Hayes extends this thinking to the collaborative and broader socio-historical aspects of her own creative practice. She situates her free improvisations and system design in collaborative research communities, pursuing joint cooperation with live electronic artists while also questioning the cultural basis of their practice. In a winning 2020 NIME conference paper co-authored with Adnan Marquez Borbon, Hayes reflects on feminist theory and cultural studies literature to expose the need for greater diversity, accessibility and interdisciplinarity within the NIME community. In a practical sense, Hayes's involvement in the participatory LLEAPP workshops present live electronic music as a path of enquiry for addressing issues of gender representation, racial imbalance and skill acquisition across a range of contexts.

Hayes observes a distinct difference between participants with and without a movement-based performance background in the LLEAP workshops. Like the other musicians, Hayes positioned herself behind a desk on the perimeter of a large multimedia room with sprung dance floor, reluctant to move in front of it. The dancers, on the other hand, were comfortable to move around the centre of the room but encountered the speakers as a barrier. Games were devised to dissolve interdisciplinary boundaries and explore the various materials available, including projections and surround sound. A shared vocabulary between practitioners evolved from this process.

The boundary-pushing exercise tested the limits of some participants, who experienced a degree of resistance due to social and historical factors. Contact improvisation, where individuals explore the fundamentals of movement in relation to others, was one of the activities used to dissolve perceived barriers. The practice needed to be adjusted to balance the personal needs and preferences of participants within the group. Combining group dynamics with individual autonomy is paramount in the enactivist view of improvisation as an effective exploratory interdisciplinary practice:

> Improvisation can be explored and studied as a highly inclusive, cross-cultural practice in which people co-create extra-musical worlds through the ongoing and reciprocal processes of exploring materials and the coordination of action. (Hayes 2019b, 459)

The framing of her collaborative research and teaching practice within the emerging field of enactive music cognition provides opportunities to recognize and accommodate the 'historically rich sensorimotor interactions of a person in the world' (Hayes 2019b, 448). Each collaborator brings a different set of cultural expectations that may make them resistant to dance-inspired practices like contact improvisation or moving freely on a sprung dance floor beyond the safety and security of tables filled with audio equipment. This same observation can be applied to acknowledging and evaluating individual creative habits. Artists who become aware of habitual patterns through such interdisciplinary collaborative activities can choose to challenge and overcome any that inhibit development and self-realization. This exploratory approach is evident in Hayes's artistic creation:

> Even when I compose now, I don't like to call myself a composer. It's about the playing and the exploration and building systems that I can feel and navigate, so it's like making systems I can play within and find something there that I'm interested in. This is why I'm so into embodiment and embodied music cognition. It's not like stuff happens in my mind and I do it. It's very much a playful, exploratory process to make the instruments, and that's really important.

The piece *Moon via Spirit* (2019) performed at the Huddersfield Contemporary Music Festival, explores Fluid Corpus Manipulation (FluCoMa), a toolkit for new sound and gesture design allowing technologically fluent artists to separate and rearrange audio based on an analysis of transients and pitch in Max/MSP and Pure Data as part of a project developed at the University of Huddersfield. Hayes plugs the drum machine input and voice into Fluid transient slice, a software-based sample slicer that detects transients, which are short high-amplitude sounds at the start of a waveform, in order to segment and reorder an audio signal. Spectral tools are used to create a glitch-like aesthetic and effects. Dense synthesis and crackling noise textures are interspersed with fragmented whispers, while warped and distorted vocal inflections trail into rapid delays as Hayes wrestles with her PS2 gaming controller to manipulate the residue of sampled vocal phrases. Experimenting with live sampling and transient detection on vocal utterances exposes new ways of reimagining vocal expression as a texture in Hayes's improvisations.

Hayes's rhythmic and textural approach to voice introduces her unique voiceprint into a noisescape shaped by twisting torso movements and improvised arm gestures. She is currently undertaking lessons with trained opera singer and noise artist, Micaela Tobin aka White Boy Scream, to expand these vocal explorations, seeking to increase the prominence of the voice in her performances. Although previously a singer in bands, Hayes acknowledges feelings of vulnerability in sharing her voice more openly in experimental and academic music settings without formal vocal training. She is keen to let this part of her practice evolve, continuing to merge truncated melodic phrases and utterances into a broader synthetic soundscape as part of her unique physical expression.

As seen in Figure 10, Hayes's performances are direct and intimate, focusing on the self-designed hybrid system and herself as a soloist, with no additional visuals to augment the scene unless she is collaborating with video artists for a

Figure 10 Lauren Sarah Hayes – live performance set-up at Lunchbox, Phoenix in 2019. Photo credit: Tobias Feltus

specific show. When asked about incorporating other forms of feedback besides touch in her work, Hayes comments that she would only embrace visuals in collaboration with an expert in the field:

> I know that I'm very gestural. I move a lot when I play. People have asked if I choreograph my feet and I say no, that's just the way that I play and I know it's a very visual thing to see, and it also gives some insight to the audience about what's going on.

When asked whether she had developed a specific gestural vocabulary, Hayes notes that she does not perceive her movement language in terms of distinct increments of time:

> You don't go on stage and perform a gesture. It's gestural. There's no discontinuous set of different movements, and I don't know if any musician plays like that really. I think that a lot of the basis for this research is founded on that idea and I think it's problematic. I've been to MOCO (International Movement and Computing Conference) a few times. We hosted it in ASU (Arizona State University) back in October and I think working with dance and movement opens it up much more. There's research looking at different ways of measuring gesture and movement.

Hayes is reluctant to accept the quantification of gesture common in NIME research, believing it is problematic to separate gestures when most musician movement is continuous. In reality, gestures are not segmented into discrete units, but rather flow together. Yet in the MOCO community, the greater inclusion of dance research in embodied interaction design opens up definitions of performer movement far more according to Hayes. This suits her more improvised and unstructured approach to gesture in performance.

Adding to her practical understandings of movement, Hayes recently pursued circus skills classes, which tested her to develop spatial and kinaesthetic awareness as her whole body navigated the air. Learning aerial arts was challenging, as was coordinating the timing of her movements to a fixed soundtrack, which was markedly different to her usual improvised mode of performance. Hayes also belongs to an interdisciplinary university department, regularly interacting and collaborating with movement experts such as choreographer and Feldenkrais practitioner, Grisha Coleman, and engineers exploring computer vision. Access to a MoCap system facilitating more sophisticated motion-tracking applications also influences the way she explores movement and thinks about future body-centred works and collaborations.

Even through Hayes does not possess long-term movement expertize, she does casually engage in yoga and utilizes kinesomatic techniques that are directed towards achieving more efficient movement by learning to feel and sense the muscles that are contracting during motion. Attending Sofia Dahl's 2018 MOCO workshop encouraging designers to develop movement skills by physically engaging with virtual reality technology was also enlightening for Hayes. In contrast, she sees the International Computer Music Conference (ICMC) and NIME communities she's part of as 'heavily focused on the engineering and design', leading Hayes to ponder, 'I wonder as musicians, those of us without backgrounds in dance and somatics, what are we really thinking about our bodies?' Hayes feels that she has learned more about how she relates to her body through workshops in somatics and circus arts than writing about sound, touch, embodiment and enactive cognition in music. She believes that these lessons from conversations and experiences among creative practitioners who perform with new movement-based electronic instruments can contribute to expanding the technical focus characterizing current research in the area.

Performing with her unique hybrid analogue and electrical gestural system has become an intuitive, subliminal experience for Hayes: 'As I get more into it, it takes me more out of the head, not really thinking about how I move, just moving in a way that feels like I'm playing the instrument and being expressive ... the more time I do it, the more fluid it becomes.' Yet Hayes also reports challenging and awkward moments on stage, like other artists of this emerging art form who were interviewed. Time was needed for integration of experience and 'embodied sense-making' post-performance to draw meaning from those difficult moments and apply the resulting lessons to subsequent works.

Although Hayes's hybrid system is deeply personal, aspects of the instrument have been incorporated in projects serving individuals with complex learning difficulties through Notts Charity in the UK. She is also interested in creating collaborative instruments in the future, 'conceiving sets of relationships between people and spaces and technologies and materials'. Hayes intends to keep participating in interdisciplinary collaborations, building on past work with dance and somatic practitioner, Jessica Rajko, in which she performed with dancers and a laser, as well as distributing haptics to the audience. She plans to continue to emphasize the potential of bespoke systems that reflect their players' cultural and historical experiences:

If I hadn't had training on piano from a young age, maybe I would be fine playing an iPad for music. Maybe I wouldn't miss that physicality. That's why I think enactive embodied histories are important, and why I don't give my system to other people, because it was like what was your experience and history of playing music and if that was on a flat screen then maybe an instrument based on that is going to make more sense, so I think that avoids those traps of generalising those things.

Hayes's emphasis on tactility in design reflects her background as a pianist, while her recent encounters with circus arts, working with students from diverse movement backgrounds, collaborating with interdisciplinary artists, and developing interactive experiences for individuals with cochlear implants and learning challenges has broadened her approaches to design and performance. This cumulative embodied history demonstrates the contribution of customized artist-designed instruments to realizing individual creativity. Moves to standardize and generalize designs may not work for some musicians. Even though Hayes assembles widely available software and commercial hardware, her unique arrangement of equipment and code sets her solo improvised performances and movement style apart from other performers. Highly developed programming skills enable her to produce unique creations and process the voice in ways that can't be replicated, immortalized in recent streaming performances. Her work provides valuable lessons in the areas of interdisciplinary solo and group improvisation and the role of tangible feedback in physically feeling sound while performing live electronic music.

9

Bent Leather Band

The Bent Leather Band draw on traditional instrumental design and embrace the notion of play in performance with movement-based instruments. Their instruments include the ornately carved LightHarp and leather-crafted bassoons inspired by Indian, Renaissance and Baroque instrument traditions. Their improvised explorations occur within the context of skilled ensemble performance, referring to their music as a digital form of Percy Grainger's free music. The Australian composer sought to create works that rejected pitch and timing conventions limiting music to scales and regular rhythmic pulses, instead experimenting with beatless rhythms, gliding tones and microtonal ideas. The Bent Leather Band incorporate free music principles in their new instrument design and performances. Founding the group in 2001, Stuart Favilla and Joanne Cannon, as keen improvisers, 'embrace Free Music as an opportunity to escape the rigid harmonic constraints of traditional pitch systems and also as a departure point for the development of new specialised musical instruments' (Favilla, Cannon & Greenwood 2005, 79). Both members of the Australian experimental electronic duo, who also refer to themselves as the children of Grainger, embrace the free music concept to move beyond the regulated pitches and rhythms of notated music and develop hyperinstruments, adapted and extended instruments equipped with electronic sensors that send data used to generate or transform sound of acoustic instruments.

Rather than endorsing dynamically changeable, disposable digital musical instruments (DMIs), Favilla and Cannon promote the potential of long-term practice for adapting to the functionality and affordances of a particular instrument. Yet they observe that few novel digital instruments evolve beyond prototype form to the next generation. Favilla and Cannon (Favilla, Cannon & Greenwood 2005, 80) define a playable instrument as 'one that essentially does not limit or inhibit the development of skill'. It inspires the performer to move and produce sound in new ways, is versatile and responsive, and allows for hours of practice and immersion. Through extensive engagement with an instrument, 'the musician develops skill and expression over time as a result of the investment of play' (Cannon & Favilla 2012, 460). For this reason, they chose to finalize their instrument designs and freeze development for a decade to shift the focus from evaluation and refinement to skill acquisition, representing a markedly different approach to many systems dominating the research of New Interfaces for Musical Expression (NIME) and International Computer Music Conference (ICMC) communities.

The group developed a suite of Meta Serpent wind controllers for live improvisation, featuring the Serpentine Bassoon, a leather-crafted meta-bassoon that can be processed through Max/MSP. It combines fixed mappings and live effects such as a signature sound of multiple delays. The Contra Monster is another instrument assembled from multiple controllers, enabling ten degrees of freedom, meaning that the body is able to move in ten different directions. The bassoons are constructed of leather, in collaboration with leather artist Garry Greenwood, and complemented by vintage bakelite knobs that evoke a 1930s aesthetic reminiscent of Grainger's era. Working as a composer and producer renowned for innovation, Favilla also designed the LightHarp using lasers and spotlights to project thirty-two light-sensor virtual strings for the performer to play. It was designed as the first Indian computer music instrument, capable of playing Indian and microtonal scales.

Acoustic instruments, and particularly ornate Indian instruments like the veena, were a significant inspiration for the Bent Leather Band instrument suite. The band collaborated with many professional instrument builders and circuit board-makers to realize their designs, investing significant money and effort to have the instruments built to the appropriate specification. They were interested not only in the construction and decoration of the instruments but also in performance techniques that develop over time, pursuing a virtuosic level of playability. The following sections explore Favilla and Cannon's unique instrumental backgrounds and experiences with the instruments before outlining their conceptualization of skill acquisition in long-term engagement with artist-designed instruments.

The following interview quotes are drawn from recorded Zoom conversations that took place in July 2020. At the time of interview, Favilla and Cannon were under strict lockdown in Melbourne, Australia, during the Covid-19 pandemic in the winter of 2020, unable to travel to work, perform publicly or pursue projects like Favilla's spatial sound library for exploring spatial audio in research and composition, and Cannon's community music projects. Although the pair stopped collaborating in 2018, their current musical activities shed light on the experimental computer music tradition that evolved in Melbourne, with its roots in Grainger's ideals of tonal freedom and irregular rhythms, moving in the same circles as influential experimental artists, improvisers and instrument makers such as violinist and composer Jon Rose, and performance artist committed to amplifying and extending the body through exoskeleton and extended arm interfaces Stelarc.

Stuart Favilla

Favilla is a composer/producer with a background in synthesizer and computer music performance. He also has a strong interest in spatial sound and classical Indian music, specifically South Indian Carnatic music. His first instrument for spatial distribution of sound was controlled by a joystick, yet the lack of visual impact made Favilla question its applicability for live work. He then built his first light harp before collaborating with violin and shakahuchi maker David Brown and computer instrument designer Robin Whittle, who developed a MIDI board for sound generation. Light sensors and a specialized controller like a theremin use magnets to detect hand movements in space, activating virtual strings and producing Indian pitch ornaments, Favilla explains:

> The way it works in Indian music is a note is actually composed from a pitch and an ornament or any embellishment, which is a really cool idea, and I thought, why don't I do that with electronic music? Why can't we have sounds on our synthesizers that are embodied with gestural elements? So that was my thing and after that I stopped composing scores and I thought there's not enough time to be codified on paper. It has to be explored. So this is about 1996 that I came to this realisation and by then I had created my first light harps from fibreglass.

The resulting LightHarp, pictured in Figure 11, allows Favilla (1996) to play in different tunings, producing dense textures, microtonal scales, portamenti and glissandi using empty-handed gestures that perform pitch-bend ornaments called gamekas with an object resembling the curves and decorative features of the traditional Indian instrument the veena. It incorporates a breath controller, pitch and modulation joystick, pressure- and position-sensitive strips, electromagnetic proximity sensor and foot-control pedals. The right hand makes attack gestures while the left hand is mapped to modification movements, often

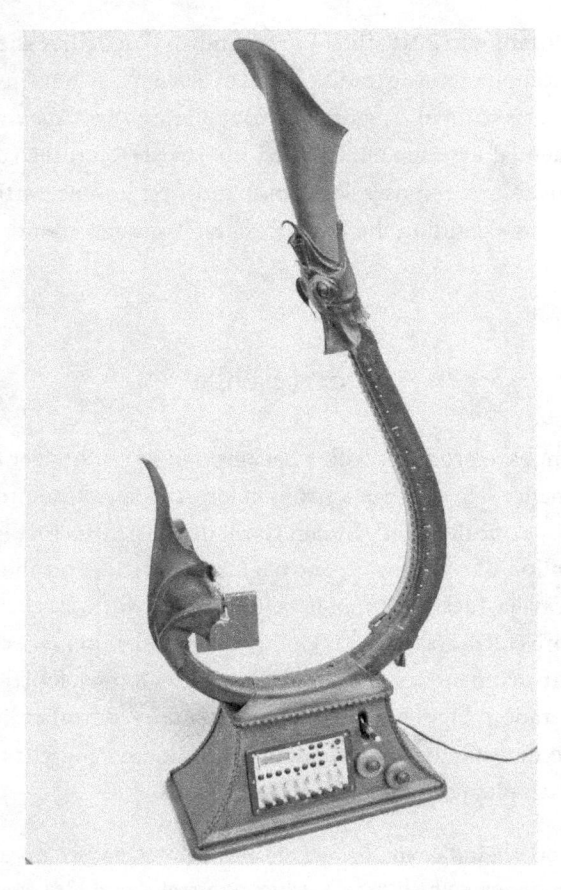

Figure 11 LightHarp showing light sensors, pressure sensors and left-facing ancillary control station including pitch-bend controller (breath controller for attack and volume control omitted)

complemented by breath control to regulate attack and dynamics. While the breath controller is generally used to control volume, light sensors regulate pitch and sample triggering. The design process signalled the start of many years spent recording, studying and sequencing body gesture as an information stream, using 'a random set of social research methodologies to try and triangulate what expression was'.

When playing the LightHarp with Cannon on the Serpentine Bassoon or Contra Monster, Favilla developed his own gestural language for controlling samples by gently caressing the virtual strings, noting 'sometimes gestures discovered their own sounds'. Commonly used sweeping hand motions became immortalized in the LightHarp's design. In collaboration with musical instrument-builder David Brown, the shape of the LightHarp was matched to the curved contours of Favilla's predominant movements:

> When we came to design the neck, we grabbed a piece of cardboard and held it against us and put a pencil in the hand and then swung the arm down in a glissandi gesture and it created the curve, and that was the shape that we took and tried to work with. So it was built around that kind of portamenti, glissandi technique.

The design is further individualized through decoration inspired by Indian deification and iconography, realized in collaboration with leather artist Greenwood, who created petals and leaves around the shape of the Yalic dragon of Indian mythology from cowhide and buffalo skin offcuts, introducing a new ornamental aesthetic uncommon in computer music, with its functional preoccupations.

Favilla mentions that after ten years of freezing development of the LightHarp and other Bent Leather Band instruments; it gave him valuable discovery moments – new breakthroughs that transcended the usual thresholds of innovation-driven performance, observing, 'There's a praxis of making and playing the instruments that's very rewarding and rich but you have to invest lots of time, unlike the violin and the piano and those kinds of instruments.' These discoveries would often yield new pieces, borne from the two-person job of experimenting with mappings between movement and sound, where one musician would play their instrument and the other would adjust parameters on the computer. Without generations of techniques and refinement behind these custom DMIs, Favilla become adept at inventing new types of performance approaches, adapting aspects of his piano and synthesizer practice:

> It does reach a point in time where there's enough going on where you can let go and just play and it was a really enjoyable time for me because the instruments would become rich in themselves through that decade of practicing and then playing and then modifying. I got to the state where I could play them and discover things every few days. A new sound, a new technique … maybe not a technique but the instrument would start to talk to you and tell you little secrets about what it could do.

At the same time, Favilla was playing the theremin, enjoying gliding swiftly through space, honing his skill in operating non-tactile instruments and experiencing 'the fact that there was no physical friction in your embodiment of sound. It allows you to play things that are so fast and so dense and so unheard of that it's really great.' The act of engaging with the theremin and LightHarp in parallel allowed Favilla to step into new paradigms of spatial performance. Researching playability also encouraged him to move more consciously and deliberately:

> I think you have more spatial awareness. I've always had a shocking posture. Your posture gets corrected or changed. You're more mindful of it. You either become a happy hunchback or an Alexander freak. But I've been reasonably lucky with my body. I play the piano every day and I haven't had any injuries in my hands as I get older. Musical instruments – there's a long tradition of them causing physical ailments as well, because there's a total impracticality in practicing twelve hours a day.

When the instrument has invisible strings like the LightHarp, kinaesthetic feedback becomes the default option for achieving nuanced expression. A refined degree of kinaesthetic awareness necessitates an advanced level of skill development where there is reduced reliance on visual feedback and attention is redirected to *feeling* the instrument through the proprioceptive and haptic senses (Cannon & Favilla 2012, 460).

Continually drawing parallels between the LightHarp and acoustic instruments, particularly those that are difficult to play such as the harp and veena, Favilla remarks, 'all the acoustic instruments have so much to teach about the things that we make and the things that we do'. The veena, for example, has a large fretboard that dominates the player's focus, making the musician more attentive to visual feedback compared to other instruments like voice, violin or piano. Favilla is keenly aware of the socio-historical significance of his relationship with the self-made instrument:

> There's a terrible dry conservatism around what an instrument should be. It's part of you. It's an extension of you. It's a reflection of your vision, and it's also another agent, another actor on the stage, like Francois Rabaud gets up with his double bass and the solo bass has a personality all of its own, and we're intrigued by that. Or someone gets up there with a platinum flute. Let's face it. The instrument is there as an actor, as a cultural protagonist, and you better think about that because there's huge potential for dialogue with your audience and instrument makers.

Favilla accentuates the role of the performer's unique voice and interpretation, even when playing conventional instruments with centuries of tradition behind them, noting that 'the real voice of an instrument emerges over time'. Musicians invent their own techniques and traditions as players. Favilla's experience with the LightHarp as a cultural protagonist reflects the acquired knowledge of expert luthiers, Indian music and his own background in spatial sound, jazz and improvisation.

Favilla is influenced not only by the traditions of jazz and Indian music but also by computer music. He makes some interesting observations about the movement from acousmatic music, delivered through loudspeakers, to the start of international conferences on NIME in 2001. The annual event and community surrounding it began as a workshop at the Association for Computing Machinery (ACM) Conference on Human Factors in Computing Systems (CHI) in 2001, splintering off from the International Computer Music Conference (ICMC) to explore human–computer interaction and gestural interfaces in greater depth. The move was spearheaded by a range of influential scholars in the field such as Marcelo Wanderley. Before that point, Favilla remembers attending conferences focused on the sound object and acousmatic works in which only a small number of women would attend. No artist researchers performed live. This 'hierarchy of patriarchal bliss', as Favilla calls it, is giving way to a broader conception of composition and computer music where 'everyone is sharing agency now with the machine and some people are making social networked collaborative performances as well, and that's the space that's really burgeoning now. That's the new frontier'. Together, Favilla and Cannon, with subsequent collaborators like jazz saxophonist Tony Hicks, explore the potential of extended instruments and DMIs in ensemble performance. Unlike the majority of artists who perform with these types of instruments in solo situations, the two musicians' cooperative approach offers new insights into the underexplored area of improvised ensemble DMI performance (Favilla & Cannon 2006, 111).

Grainger's influence looms large in their work. The Bent Leather Band recorded live at the Grainger Museum in Melbourne, updating the concept of free music the composer espoused in the 1930s. The notion of pitchless and beatless music lends itself to the physically inspired movement curves of gestural music, as Grainger's theoretical free music pieces such as *Free Music No. 1*, traditionally scored for four theremins, attest. In reflecting on this musical legacy, Favilla considers the contradiction between Grainger's free music machines and human performances of free music. He notes that Grainger expressed frustration with the shortcomings of individual interpretations by players in theremin ensembles assembled to perform his free music compositions:

Grainger's vision was originally to have it all played on theremins, but he couldn't get people to play it how he drew it. He didn't like people interpreting his music … and he then built all these machines – this totalitarian rendering of his free music, which the composers are drawn to because it fits their paradigm. It fits the paradigm of the great patriarchal composer spitting out the truth that you must sit down and listen to, right? I reject that because I say if you want to have that just have Ableton Live or what's the point of connecting with humans?

Grainger's machine-based approach to free music contradicts his own loose timings and radically differing piano performances of Edvard Grieg and original repertoire. For Favilla, the role of the musician as an interpreter of a musical work is central to true free music principles, commenting on the expansion of participatory music in recent times:

Free music is always unsettling for computer musicians and in particular composers. Composition taught at universities is all about organising sound, I guess. But it's still a hierarchy of patriarchal bliss and it's not up-to-date. Being up-to-date is multi-agent, it's collaborative, it's working with a whole bunch of people being in the background, not letting you go forward too much. We're all part of these multi-agent systems.

Favilla channels these performance realizations into his work as a design researcher, reflecting on the link between gesture-based DMIs and key innovations like Douglas Engelbart's invention of the computer mouse, which represents a human machine augmentation approach: 'The mouse created this new spatial domain inside the processing language, and gestural musical instruments do the same thing.' Favilla has merged a range of influences in his overall approach to design, including Brian Shackel's (1986) usability model and the idea of playability drawn from computer game design, encompassing ideas like viscerality, naturalness and quality of feedback (Favilla, Cannon & Greenwood 2005, 80), as he explains in the interview:

The idea of a visceral instrument is a really interesting thing and I looked at the human nervous system and I taught myself EEG and looked at measuring synaptic latencies. And I realised the time it took for your central nervous system to communicate something to your foot was quite a lot longer than the things that went off the top of your spine like your arms. So I made that a big part of Bent Leather design – that you would use the fingers. You use the strong fingers first and then you would break out to breath and parts of the face rather than going to the feet cause the feet are going to take forever to learn how to play.

Having been in the field for over thirty years, Favilla has since transferred his insights to designing musical applications and games for patients with dementia and vision impairments, as well as lecturing in software and interaction design. He acknowledges the financial pressures on musicians in this area over the long term, particularly due to the need to contribute personal funds and effort to continue their experimental performance practice and unique instrumental design and maintenance while maintaining a consistent income. Like many computer musicians that find themselves in higher education and technological fields, Favilla has accumulated highly transferrable skills that can be applied to related fields such as health and industrial audio applications.

Joanne Cannon

Joanne Cannon is a composer and professional orchestral musician also trained in jazz and improvisation. Cannon's experience of being channelled into playing the bassoon rather than her preferred instrument, the trombone, echoes research by Lucy Green (1997, 176) that establishes how adolescent girls in the 1970s were diverted away from loud and electronic instruments. Yet Cannon found another route to making sounds with unconventional instruments capable of amplified screeches and rumbles that would contradict gender conventions associating femininity with softness and subtlety. Cannon's aim to propel the bassoon beyond the orchestral context and enhance improvisation spurred her to extend the bassoon by attaching sensors and using effects pedals to control digital signal processing. Her early efforts proved unwieldy due to the upright and condensed design of the bassoon, ultimately leading to the development of the extended digital bassoons, Serpentine Bassoon and Contra Monster, to access novel, abstract sounds.

Unlike their traditional ancestors, both instruments are equipped with sensors and controllers including joysticks and dials to perform processing within Max/ MSP. The Serpentine Bassoon maintains the length and the mouthpiece of the acoustic bassoon, with eight open tone holes for pitch control or sensor activation, whereas in the bulkier Contra Monster the tone holes are replaced by force sensitive pads, operated in conjunction with condenser microphones and multiple controllers (Favilla, Cannon & Greenwood 2005, 82). According to Cannon, 'Making a dedicated instrument that still used a crook and a double reed that was still bassoon-esque, but got rid of all of the keys and just had open holes and then

other space for controllers and sensors we added on – that was the idea.' Inspired
by the decorative instruments of Baroque and Indian music traditions, Cannon
collaborated with Favilla and Greenwood to fit the limp form of the serpent-
shaped leather to her physical dimensions. No longer confined to a specific pitch
range, the instruments can reach impossibly high and low frequency extremes,
which Cannon uses to increase her sonic palette during improvisation:

> I think of them as being timbre-based instruments that have an acoustic sound
> that can play all of these amazingly wild things, because they're processing
> digital sound and they're also connected to speakers. When you do that the
> musical language changes because you're not just this acoustic solo entity
> anymore emanating from space. You've got speakers there and it could be more
> than just two. It could be a whole field of them – spatial sound and sonics, and so
> your musical vocabulary is a lot bigger and you can start then to improvise with
> yourself even, and layer parts as well.

Cannon also often plays the acoustic bassoon alongside the Serpentine Bassoon
and Contra Monster, the latter of which is pictured next to the LightHarp in
Figure 12. The bassoon is amplified to expose breath sounds and mechanical
clicks and clacks that are usually inaudible. Cannon draws on her experience
with the extended instruments to expand her understanding of the original
instrument beyond classical repertoire:

> You don't necessarily know what's going to happen, but it evolves and it's very
> conversational. Being able to glide and playing all of those things has definitely
> affected how I play the acoustic bassoon, and quite often when you start to get
> into playing and you forget about the interface and what's happening, it becomes
> gestural in terms of what your fingers are doing. You know approximately
> whereabouts it's heading but especially with the digital instruments, it could
> also go anywhere, just through the nature of the sound processing and how
> you set up all the effects you're going to do then how it'll work together. That's
> really exciting in terms of an improvisation, because you get to work with what
> happens and you start to shape it.

Cannon often enters into Grainger territory, embracing gliding tones and
dramatic timbral variations 'with an experimental instrumental or digital
musical instrument, that was made specifically to play everything else that the
bassoon necessarily couldn't play, and so you entered the world of timbre and
gliding pitch, and everything else that technology does really well'. The speakers
also become part of the instruments, as do the unique materials used in their

Figure 12 LightHarp and Contra Monster on stage prior to Bent Leather Band show at Liszt House, Budapest

construction: 'The leather has these creaks to it and then blowing into it you can also make it crackly and hear the air.'

Yet merging with the instruments did not occur overnight, Cannon observes: 'Making the physical instrument is one thing, but then working out what I'm going to do with it is another thing.' Cannon's concept, investment of play, frames her practice with the extended bassoons. Much time is spent refining her performance technique and selecting suitable controllers for specific parameter changes. She also leverages her pre-existing instrumental skills when performing with the digital instruments:

> I'm still using all those muscles in my fingers and that dexterity that I've developed over a really long time. When you're a bassoonist, you become really good at using your thumbs. You become good at what you practice. There were also years spent playing and developing embouchure and breathing. It would be silly to just throw all of that stuff out and just start again. There are a lot of researchers in NIME exploring technology and being conference based it does drive the new interface thing too. I think to be good at something, stick with it and build on it.

This immersion in a finished instrument design enables Cannon to be in the moment during improvisation, without having to learn a different type of fingering or breath control. Liberated from the usual stylistic conventions of the classical bassoon set by orchestral repertoire, Cannon is free to adopt an experimental approach rather than aiming for perfection and virtuosic interpretation typical of existing repertoire.

Both the Contra Monster and Serpentine Bassoon conform to Cannon's body shape, incorporating her playing techniques as a skilled bassoonist. Whereas she sits while playing the smaller Serpent Bassoon, part of the bell of the Contra Monster sits over her shoulder as she stands, allowing her to feel the weight of the heavy instrument as she moves with it. This movement changes what she plays and how she plays it. Cannon has now reached a stage where she is at one with the instruments:

> When I'm playing those instruments I don't feel like I'm separate. I feel like I am with it. That's something that develops over time. It's not something that happens immediately. When you first start something, getting your head around something, you're still coming to terms with it and how you handle it. What do you do? Where's the best place to go? How do you fit around it? When I play now, it's not something that I feel like I have to think too much about it in terms of where my hands go, how my mouth fits around the reed or where it is. It just is there. And over time, I feel that all of those physical movements that I do make, they're part of the sound as well.

Cannon is keenly aware of the places where her body holds tension in performance, and how that might affect her playing and posture in the long term: 'With the bassoon and the other extended bassoons, I know how to sit to be relaxed. I know if I'm holding tension somewhere, then afterwards I'm going to be really sore.' To alleviate strain caused by the weight of the heavy acoustic bassoon, Cannon places a spike underneath it, allowing her to relax her shoulders. It is a form of self-preservation and nurturing – keeping the body in optimum condition is as crucial as maintaining the instrument itself. This process is not necessarily conscious: 'When you walk you're adjusting and you're not thinking about it. The instrument's very much part of the body, and I guess also blowing too. Plus there's the amount of breath that goes through at any one time and whether you want to stop it or whatever you want to do.' Similarly, breath control blends preconscious and deliberate movements, including articulating musical phrasing through inhalations and exhalations, bending sounds, talking through the instruments and exploring the flexibility of the double reed. Cannon describes her immersion in the experience: 'It feels almost like I'm dancing sometimes. It feels like there's a lot of movement and dance and gesture all tied up with the music.'

Cannon is currently working on a number of recordings for the acoustic and digital bassoons. These projects include collaborations with other improvisers centred around her poetry and the first recorded version of Bach's *Viola Da Gamba* sonatas for acoustic bassoon. She is also planning to write a book about the multiphonic potential of the bassoon. She continues to feature the meta-instruments in her performance practice:

> You're creating all these timbres and to me I feel like I'm just playing jazz even though it's wild, weird, raucous, abstract, whatever, but in my head, there's still that essence of jazz and improvising. If I'm playing with someone else there's all that listening and interaction. There's still that regard for form in what's happening. There's still a regard for repetition in the sense of developing what's going to happen. All of those musical things are still there and they're the same things that you get in dance when you're improvising and dancing. Harmony becomes timbre. Pitch is there but it's movable, gliding.

When playing live with technology, Cannon acknowledges that technical issues can disrupt a performance, but the ability to overcome these unexpected glitches contributes to the mastery of these bespoke instruments. Unlike the exacting nature of interpreting baroque repertoire, Cannon characterizes her approach to the digital bassoons as exploratory and unconfined to a predictable outcome, leading her to conclude: 'I think when you've finished something or if you think that you've come to the end of something, that's not creation. Creation is ongoing.'

The investment of play

Cannon's concept of the investment of play was borne from an understanding that creation is continuous. It is a lifelong process. A musician can never achieve complete mastery of an instrument. She disputes the notion that 10,000 hours of practice are needed to master a skill or become an expert, as outlined in the book *Outliers* by Malcom Gladwell (2008). More relevant are how those hours are spent and the types of skills a musician develops over the longer term, Cannon contends.

Cannon and Favilla (2012) note the lack of longitudinal studies demonstrating the role of hours of play in instrumental skill acquisition. They fill this gap with a focus on the role of long-term play on DMI design and evolution, analysing the role of skill development in changing an instrument's affordances. This research contributes to their definition of a playable instrument – one that is palpable and that a musician can practice for hours to hone their skills (Favilla & Cannon 2006). The Bent Leather Band deliberately froze development for six years, playing and perfecting their meta-instruments after a decade's worth of development. They also analysed eight channels of sensor data from audiovisual recordings during preparation for ensemble concerts dating back to 2008, completing a sixteen-year study of the band's instrumental practice and development. Since the mid-1990s, they estimate their investment in play as a duo to be four thousand hours (Cannon & Favilla 2012). They argue that expression is directly linked to the invested play uncovering the latent affordances of an instrument.

Yet the process of preserving the instruments in a fixed state is hampered by perpetual software updates, Favilla observes. These challenges must be overcome by periodically pausing development in digital instruments, as Waisvisz (2000) recommends, to allow the performer to refine their techniques and body schema when playing an instrument. Favilla laments the rise in usability that diminished the role of skill in computer music by promoting ease of use in design. Just like acoustic instruments, DMIs require dedicated practice, contradicting earlier moves to reduce effort and make instruments that are instantly playable. Investment of play encourages fluidity of action and refined musical expression, while encouraging the emergence of personal style when playing customized instruments (Cannon & Favilla 2012, 466). This exploratory method has parallels with soma-based design approaches adopted by the next two artists, Garth Paine and Mark Coniglio.

10

Design reflections

The next two interviews are with artists who provide insights into designing systems across a range of art forms and developing software for other artists. Composer Mark Coniglio invented Isadora, a graphical programming environment for artists, designers and technicians to manipulate digital video and sound. Garth Paine produces solo and large-scale audiovisual and collaborative works, studying the efficacy of digital musical instruments (DMIs) and their links to the somatic dimension. Both artists' works are interdisciplinary, founded on collaborations with dancers and visual artists that inform their compositional decisions and technological innovations.

Coniglio's programming and compositional projects foreground the body, reflecting the rhythms and embodied skills of dancers, choreographers and theatrical performers, while exploring relationships between sound, light and movement. He brings unique understandings of body movement to the development of Isadora, informed by regular interactions with dancers and participation in movement exercises as part of the Troika Ranch dance company he co-founded with Dawn Stopiello. These experiences were integral to the early development of Isadora software. Coniglio continues to incorporate feedback from users in the program's features, leading a vibrant community of creators and performers experimenting with movement expression through multiple forms of media.

Unlike the other artists interviewed, Paine does not favour a singular system, but instead uses a range of commercially available sensors and control surfaces in live performance including Nintendo WiiMotes and the Karlax, an instrument that captures gestures of the upper body. He also hacks on projects like HappyBrackets, which promotes creative coding initiatives through multiple remote devices such as Raspberry Pis. The concept of an interactive system that is absorbed into the body schema is at the heart of Garth Paine's (2015) techno-somatic design approach. He defines the techno-somatic dimension as 'the "feel" of an instrument, formed through both somatosensory feedback and listening, representing the cognitive map a performer develops in order to play an instrument, the technique, and how the instrument responds under different circumstances' (Paine 2015, 84). His conceptualization of the 'techno-somatic dimension' encapsulates the space that resides between the body and the instrument (Paine 2015).

Paine embraces somaesthetic approaches to listening, composition and design. Rather than treating the body as an external object to be observed and analysed, he adopts Shusterman's perspective of the body as central to lived experience, incorporating techniques for observing, documenting and analysing felt experience in his work. Ethnomusicologist Anna Tarvainen notes that somaesthetics differs from other body-based philosophical approaches in that it calls on the researcher to engage in bodily activities and somatic practices directly (Tarvainen 2019, 9). For Paine, cultivating awareness of the body in relation to instrumental performance and listening is an ongoing process, represented in a range of projects including investigations of musicians' instrumental experiences and listening projects that present the acoustic ecologies of American Southwestern desert environments.

The final part of this section analyses my personal performance and recording experiences associated with creating the works *Intangible Spaces* (2018), *Magnetic Springs* (2019) and a live album capturing an electronic ensemble performance, *Youbeme* (2021). The pieces reflect a strong need to experience music viscerally, particularly rhythm. Like Hayes, Cannon and Favilla, improvisation is a significant component of my performance and design practice. Gestural systems not only offer the ability to move to an existing rhythm but also to shape phrases, accents and impose idiosyncratic body rhythms on the pace and momentum of a live composition. The body informs the design of a customized non-tactile system that has become my main instrument in recent years. Ongoing physical engagement with the instrument throughout the design process helped me develop a grounded, embodied understanding of how movement works in relation to sound generation and processing within the customized interaction environment.

11

Mark Coniglio

Mark Coniglio is a composer, media artist and programmer known for large-scale collaborative multimedia works that merge music, dance, theatre and interactive video. He blends these complementary skills to create Isadora, audiovisual software for performers, designers and VJs, including high-profile users such as Pamela Z and director Francis Ford Coppola. The interactive media program, named after contemporary dance innovator, Isadora Duncan, is a visual programming environment facilitating real-time digital video and support of musical instrument digital interface (MIDI) and open sound control (OSC). Coniglio co-founded Troika Ranch, an arts organization dedicated to the exploration of the moving body in relation to technology with choreographer Dawn Stopiello in 1994. The company produces performances, interactive installations and digital films that engage with contemporary themes, provoking reflection on the evolution of physical habits and the integration of body and machine.

Coniglio began developing Isadora in 1999 as part of his work with Troika Ranch, building on an earlier major software innovation, MidiDancer, a bodysuit fitted with plastic fibres that measure and translate the flexion and extension of the main joints into code used to manipulate digital sound and video. Until then, his focus had been on creating instruments for dancers' bodies. Isadora signalled a shift towards video processing to complement sound control. The software's development is influenced by the unique choreographic and spatial knowledge of professional dancers and choreographers, many with limited technical experience, through a series of Live Eye workshops teaching Interactor, a MIDI-based program Coniglio developed with Morton Subotnick, and video processing software, Ima/gine. Coniglio regarded this period as a valuable beta testing phase that guided the early development of Isadora, as he explains in a Zoom conversation in July 2020:

> Literally we would teach the workshop during the day with these dancers and
> I would witness them fail at some part of the program. I'd go home, fix it, compile

a new version and bring it in for them all to install on their computer the next day, so we were really in situ with these people who were choreographers and dancers who think spatially. They don't necessarily think like programmers.

Using the responses and feedback of dancers and choreographers as a basis for software improvements embedded embodied understandings in the design. The workshops also inspired enhancements to the graphical user influence (GUI) directed at creating a friendlier, more accessible interface for users with strong physical abilities who do not necessarily possess advanced technical skills or a programming background.

Coniglio's dream, which he shared with Stopiello, was to 'make a software where any choreographer can make an interactive media piece without necessarily being trained for it':

I absolutely want to see everyone make amazing work that makes your heart move, and throws you off balance, lets you reformulate your relationship to the world and ask questions about who you are in this world that we live in and how you relate to it. I want every single person who is an artist to be able to do that, and if I can give of myself a tool to make that become true, then that's something that actually gives me pleasure and contentment to know that I facilitated that.

To achieve this design aim, Coniglio addresses the barrier of limited technical knowledge for artists wanting to actively interact with media in the program's layout:

Computers in the end always force you down a certain road of thinking in general. If you don't feel comfortable you're going to go to a certain point where you're going to need to bring in someone like me who has that very intimate relationship with that kind of thinking who can take it a bit further. There's a certain point where you can't go beyond if you don't have that sensibility.

Isadora was designed to be accessible to artists from a range of backgrounds, reducing reliance on technical collaborators. Users can blend audio, video and other media for performance and installation works without prior programming experience. In its latest iteration, the software allows for multiple camera inputs, enabling users to capture input from a variety of sensors. Coniglio sets out to create software that welcomes users with visual hints and clear naming to navigate the program, in contrast with other node-based software like Max/ MSP or Pure Data, which open with blank, empty patches. Unlike other visual programming environments for processing data, Isadora has clearly marked actors, which are like virtual objects or devices that perform a single function from a Video Mixer to Video Delay. The actors are assembled in a scene where the user can process one or more streams of data. In Coniglio's view, 'having an icon for every actor that gives you a sense graphically, with a picture of what it does, having every input where the value is visible, and the name of that parameter is visible' familiarizes the user with Isadora's visual programming environment. He believes this invites the user in, welcoming them into a new setting where the rules are clearly mapped out, as opposed to software that compels users to enter the programmers' own worlds and ways of working.

This approach illustrates a strong theme of equity that runs through Coniglio's work with Isadora. The community that has evolved around the software are welcoming and supportive. Users freely share details about performances and installations created with the software and exchange patches at practical workshops and online. Evidence of trolling and other forms of online abuse on many other software user forums is notably absent from the Isadora TroikaTronix forum. Coniglio ponders whether this open atmosphere is perhaps partially attributable to the large proportion of female users, as well as the presence of artists and designers from varied backgrounds. Coniglio and the Isadora community are currently broadening their focus to find strategies for expanding this theme of inclusivity by diversifying its membership further.

Although it was first developed to meet the artistic needs of his work with Troika Ranch, the software's features 'just happened to intersect with the public', according to Coniglio. Isadora is flexible and open enough for the user to customize and express their own unique aesthetic. In recent years, theatre designers have discovered new applications for the software that Coniglio could never have envisaged, providing a more cost-effective solution to other exorbitantly priced media server software on the market. Yet technology is not the endpoint as far as Coniglio is concerned. Although many real-time video generation and processing programs present marketing material to seduce potential users with slick, three-dimensional (3D) visual content generated using flashy particle systems, a technique for recreating random elements of natural phenomena like smoke, water and snow in real time, the technology itself offers only short-term gratification, Coniglio asserts. Art instead should aim beyond first impressions to put people off balance and upset their complacency, he contends.

Coniglio considers himself fortunate enough to own a computer from an early age, permitting him to learn programming at his own pace. He was employed professionally as a programmer from the age of sixteen. He also worked as a music producer before studying with Subotnick, signalling the start of a productive period of combining his passion for music with programming. Coniglio developed software to measure the tempo of a conductor with an Air Drum sensor for Subotnick's piece for fourteen-piece chamber orchestra, electronic instruments and MIDI score, *The Key to Songs* (1986). This was Coniglio's first foray into interpreting and translating human movement to apply variations to a pre-composed MIDI score: 'The tempo is following the beat he's conducting, so that the flexibility, timing and dynamics that a conductor can impose on the group can also be imposed on the MIDI score as well,' Coniglio explains. This experience solidified his belief that 'human beings are the most powerful source of information if they're in the moment reacting to the audience and their own feeling that day', leading to an ongoing fascination with the expressiveness inherent in the human body and the heightened physical skills of dancers:

> They understand the audience. They're in the moment. If they have the flexibility and they're not performing to a CD, which imposes the structure and timing on them, it'll be a little bit different every day they're in it. That's what I wanted to allow with Isadora and with the sensor – is to take the music that I created for them and to reinterpret it in that way so their instincts as a live performer will be part of the mix. I consider that part of the lineage I got from Mort in terms of what he was doing with electronic music.

Enabling dancers to impose their own intuitive timing and movement phrasing on a work expanded the potential for unique performances that would reflect their prevailing thoughts, feelings, energetic states and their novel interactions with each new audience. The performer was no longer sublimated to a predesigned composition or choreography, leaving greater room for individual expression. Redirecting control of rhythm and pacing to the performer also solved the problem of precise and mechanical timings created by the introduction of MIDI. Coniglio could then channel the physical input and interpretations of the dancers into the realization of his own compositions.

In his collaborations with Troika Ranch, Coniglio reports that music was often composed in the final stages of creating a work, after visiting Dawn and the other dancers involved in the choreography:

> Normally, most choreographers would have a piece of music completed or at least halfway completed and then start choreographing to it using the music as inspiration. Dawn and I worked the opposite way because I would watch the

choreography and go home and respond to what I saw and write music, so the dance usually came first and then the music came as a secondary part of it.

Movement is thus foregrounded in the process of creating a Troika Ranch work, followed closely by sonic elements:

> The music came very last most of the time, even after visuals, because the dance and the visuals are getting created and then the music would eventually come along. It was very luxurious for me as a composer to work in that way.

Coniglio describes this interdisciplinary work as an experimental and evolutionary process in which software patches, or seeds of ideas, are explored physically:

> Before we started working with the dancers, I would make an Isadora patch. When we were using Isadora, with 30 or 40 different things we thought might be interesting, and Dawn and I would go into the studio together alone, and she would put on whatever sensors we were using, and we would try those 30 or 40 ideas. We'd end up with ten by the end that we thought were promising, that had something going for them, and also through that process we started to already refine the actual interactive system, but once we got it to that point, then we would bring the dancers in the company and let them be in it.

The primary motivation of this creative design workflow was to discover novel movement choreographies and define links between action, sound and visuals that evoked potent and satisfying feelings. Coniglio remarked that the dancers who worked in the company 'understood we were giving them an instrument and they had lots of ideas about how to use those. So it became a collaboration between all of us as we saw what the possibilities were. At that point it was about: what can we do with this? What feels good? What makes me move in a different way?'

The combined music, video and dance components are driven by a central theme or concept decided by Coniglio and Stopiello, from exploring the relationship between humans and animals in *16 [R]evolutions* to dissolving boundaries between performer and audience in *SWARM*, which incorporates audience engagement, resulting in unpredictable collective action. The large-scale work, *16 [R]evolutions* employs motion-tracking technology to detect dancer movements, which influence interactive 3D visual imagery, depicted in Figure 13. The opening interactive image is a circle controlled by a solo dancer's face, which changes in size before eventually dissolving into amphibian-like bubbles then diamonds, matched by aggressive sound. A stark vertical line theme shows the transition from animal intuition to human intellect, Coniglio outlines:

> A straight line goes all the way to the top of the screen and all the way to the bottom. A perfectly straight line does not exist in nature. That's not actually part of the natural world. That's part of the concept of geometry and mathematics. So in the first three and a half minute improvisation, because the dancer did improvise it every night, the structure of these images was the same. It's like the beginning of an opera. It's the little thing that tells you everything you're about to see. It's the whole piece. It's the development from being an animal into being an intellectual creature. We do that in three and a half minutes and in the entire piece we start over from the beginning as animals and make that journey again. Because in the end, the point of that piece is can we integrate those two parts of ourselves and is it important that we integrate those two parts of ourselves?

Music, dance and visuals are all linked in service of the central theme. Coniglio aims to lock sound and images together, stamping each work with this unique interdisciplinary style. This process is informed by direct experience with dance improvisation and exercises, stimulating a unique understanding of the needs of

Figure 13 Lucia Tong, Daniel Suominen and Robert Clark in *16 [R]evolutions* by Troika Ranch. Photo credit: Richard Termine

a robust and stable system that can accommodate a dancer's need to warm up and navigate spatial parameters without encumberment.

When Coniglio himself had to dance in a production, he found his spatial skills were tested:

> I was in one of the pieces. I had to learn some choreography. That was one of the hardest things that I ever did because I don't think my brain works that way. I have a really hard time working in 3D stuff on the computer even. That's also spatial. I have a really hard time with that actually. It's difficult for me to conceptualise.

Yet acquiring somatic skills enriched Coniglio's practice as a programmer:

> Hybridisation has been so important for me personally. I'd definitely recommend it for everyone. I did warm up with the company. I'm not a dancer but I still warmed up with them often. We did a lot of physical exercises that we learned and developed growing out of what we learned from a guy called Scott Calman in Los Angeles, who was an improv director. I did all that stuff with them. I performed in pieces and actually had to learn the choreography. All of that informs the way I create the code.

A key impetus for Coniglio and Stopiello was to allow technology to transform choreographic invention. Troika Ranch works like *SWARM* and *loopdiver* examine the effects of technology on the body and how it moves. *SWARM* is a collaborative composition exploring the relationships between bodies in space and the potential of technology to bring humans closer together. It highlights the micro-movements humans make when using touchpads and mobile phones, such as pinch and swipe gestures, which contribute to a new touchscreen gestural vocabulary. The choreography for *loopdiver* capitalizes on this emerging movement language, exploring the looping function of computer-based media manipulation and the jerky movements of interrupted video streams and animated gifs. Seen in Figure 14, the production features live performance of machine-inspired repetitive movement patterns. Dancers perform and embody software editing functions, signifying how prevailing technologies influence human behaviour, often beyond conscious awareness. Coniglio observes that the piece created a novel choreographic approach in terms of imposing a looping strategy on human dancers inspired directly by technology.

For future Isadora releases, Coniglio is conducting an investigation into the possibilities of incorporating multichannel sound. Currently based in Europe, he

Figure 14 Troika Ranch performs *loopdiver*. Photo credit: Alexandra Matzke

continues to blend collaborative artistic and design work directed at artists, one activity informing the other. He is pondering the creation of a telematic piece for four dancers in different parts of the world to accommodate an international shift to remote working since the Covid-19 pandemic began. Coniglio's focus on interactive design and technology continues to prioritize the rhythms and intuitive movements of the human performer, overriding strict machine time. His artistic focus is shaped by a long-term involvement with contemporary dance and an admiration for 'people expressing emotions and even abstractions with their body'. From the mid-1990s Coniglio has been dedicated to 'creating instruments for the dancers' bodies':

> I provide interactive control to the performers as a way of imposing the chaos of the organic on to the fixed nature of the electronic, ensuring that the digital materials remain as fluid and alive as the performers themselves. (Coniglio 2005, 8)

This is a trajectory Coniglio traced early in his career with mentor Subotnick and continues with his regular contributions to Isadora's Troikatronix community and the development of a flexible software platform that is affordable to artists on modest freelance incomes and small- to medium-scale production companies wanting to explore links between movement, sound, lighting and video. Isadora software integrates the physical experiences and intuitive rhythms of dancers and other artists exploring movement with electronic processes and media, inspiring new choreographies and relationships between digital and human data.

12

Garth Paine

Composer and researcher Garth Paine places the sensing, feeling body at the centre of his gestural musical performance and design pursuits. Paine's works range from solo live electronic performances to interactive dance works and public art installations. He regularly works with interdisciplinary teams including somatic practitioners and designers, exploring connections between the performer's body and musical instruments in an effort to understand how the two interrelate through skills and sensitivities. His approach to interactive and responsive systems is relational and somatic, recognizing the possibilities of these emerging technologies as they intersect with the experiences and biology of the human player. His individual philosophy is closely connected to soma-based design, studying how the ergonomics of an interface or device intersect with a performer's abilities and understanding.

Paine (2015) introduced the idea of the techno-somatic dimension to describe the interaction between player and instrument. He treats the relationship between performer and their instrument as a distinct realm, whether it be directed through the body or is external to it, as he describes in a recoded online conversation via Zoom in October 2020:

> This space between me and the instrument, which I refer to as the techno-somatic, is a subtle body embodied relationship, or set of relationships ... that space between the body or between the embodied sensibility and the instrument itself as an object – as a thing that we are using to create music.

When Paine describes the techno-somatic space, he is referring to an intermediary dimension connecting the body and instrument:

> We're not playing the instrument directly. We're playing the instrument through a layer which is our evolved sensibility to the instrument – our somatic relationships to it – our perceived mapping of excitation as an activity from the body into the instrument and its response to that.

A techno-somatic design approach moves beyond the disparate methods of individual instrument makers into a higher-order analysis that conceptualizes the fit between individuals and instruments, recognizing the nuanced and shifting relations that evolve between the two. It offers a flexible method of interpreting how musicians of different body types and movement preferences adapt to both traditional acoustic and novel digital musical instruments, counteracting the individualized design approaches as well as the technological and empirical focus that characterizes much NIME (New Interfaces for Musical Expression) research, Paine asserts:

> We build idiomatic approaches to performance which builds also with it literature and that literature in terms of music becomes something we share so we build a body of practice and we have communities of practitioners that feed the evolution of practice, which hasn't happened so much in the NIME community cause people are more focused on having idiosyncratic, individualized responses to the technology.

Musicians incorporate their unique histories and physical experience in instrumental practice, reflected in muscle and general body memory. Paine highlights the benefits of recognizing the intermediary layer between the body and instrument as it shifts the focus from instrumental object development to understanding 'the energy that's within the body and the way in which one reflects or illustrates emotional properties through the body', he outlines.

The techno-somatic dimension exposes the need to understand how musicians relate to particular technologies and media, whether they be traditional instruments with clear causal effects between action and sound, or digital systems, where the connections between gestural excitation and sonification are less obvious. Performance strategies designed for either instrument type can include certain fingerings or breathing techniques, linking the body to the constructed object or system in a range of ways:

> The dimension itself can be considered porous, where the porosity is a product of varying (hopefully increasing) technical proficiency, the context of use (performance or rehearsal), collaborative engagement and variation in body form from individual to individual and as a product of ageing, producing possible variation in somatosensory and listening phenomena. (Paine 2015, 85)

This theorization offers new understandings of how musicians invent novel types of interactions with customized instruments. Uniting technical design priorities with attention to somatic aspects that address 'how the instrument fits the body, how nuanced the relationships are between exertion of the body (breath, gesture) and the resultant sound' (Paine 2015, 88) sheds light on how musicians from a range of genres engage with gestural systems, exposing 'the subtleties of differentiation in the kind of agency and embodiment that is afforded through engaging with digital systems, and how that allows us to gain certain freedoms – certain kind of vectors of change are always alive and always active in ways that they're not typically in more traditional practice', Paine notes.

Broadening the techno-somatic dimension beyond the performer and instrument and their own body, Paine's model also encompasses the audience: 'I see the whole thing constantly as a bigger picture kind of embodied space. Any performance action that we take, in public or otherwise, is this much bigger picture of enmeshed sets of relationships that are very much highly multidimensional.' The idea of embodiment extending beyond the individual to their surrounding environment is prominent in Paine's live electronic music performances and interdisciplinary works. In his immersive 360° surround sound work *Future Perfect*, composed with field recordings captured in urban parks throughout Paris and Karlsruhe and the deserts of Arizona, audience members are able to augment the score using smartphones to control interactive sound synthesis so that the sounds appear to be coming from different points in the space and shift from one person to another, linking them to the ecosystem of the performance and installation:

> You can see the flow through of all of these ideas into *Future Perfect*, where not only are they in a high order ambisonic sound field, but they've got a VR film being projected around them and their personal cell phone becomes a speaker in the work. Then everything becomes a kind of ecosystem. The work itself becomes this dynamic ecosystem. And so the performance becomes this thing of bringing that to life, and not just me embodying and having agency, but everybody who's in the performance embodying the work and having agency in it. So there's no delineation or no binary of performer/audience.

This approach unites Paine's two interests of embodiment and acoustic ecology. The work sets out to 'engage the audience as an ecosystem'. Audience members are invited to step out of the passive observer role and actively contribute to and transmit sound through loudspeakers distributed in the space via networked smartphones, as shown in Figure 15, acting as agents in the delivery of the ambisonic composition. The sound spatialization and interactive elements of the work were developed at IRCAM (Institute for Research and Coordination in Acoustics/Music) in Paris,

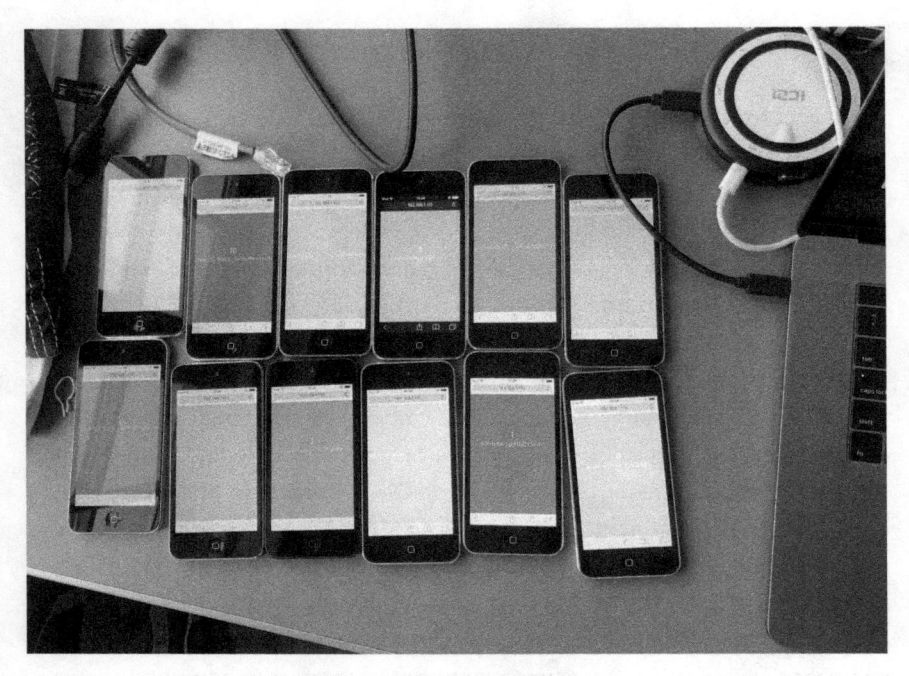

Figure 15 Networked smartphones performing real-time sound synthesis for *Future Perfect*, using the Active Listener system, developed by Garth Paine and the Sound Music Movement Interaction team at IRCAM, Paris in 2018

pictured in Figure 16, during an artist/researcher residency in association with the ZKM Centre for Art and Media in Karlsruhe, Germany.

A related series of virtual reality (VR) experiences, *EcoRift,* allow participants to enter into remote pristine natural environments, aided by 360° photography, video and matching 3D ambisonic sound, experiencing a new type of interaction with environments usually out of reach (Feisst & Paine 2020, 220):

> The quality of their haptic/sensory engagement with the interface (their directed gaze), the feedback loop formed by somatosensory and listening phenomena, forms a techno-somatic dimension that informs and shapes both the elements it links (user and technological mediation) and, through an awareness and sensitivity to the potential and latent agency of the relationship, between soma and techne, a materiality can be defined that is fluid, viscous, and porous, inviting the user to be present, to suspend disbelief and to feel free to venture across the globe and revel in the wonders and value of these often overlooked natural environments.

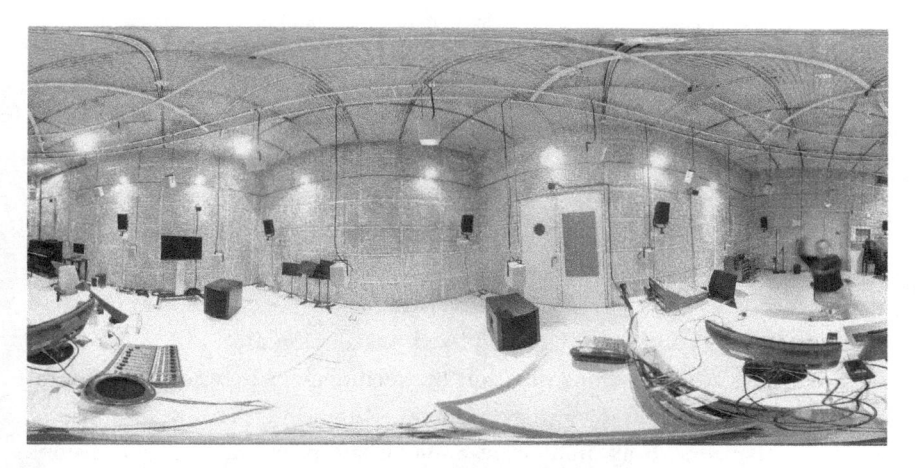

Figure 16 Dr Garth Paine composing the ambisonic score for *Future Perfect* in Studio 1, IRCAM, Paris

The VR experiences of *EcoRift* promote a new type of embodied engagement in which participants can experience natural environments and respond viscerally and somatically without physically occupying the actual spaces. Head tracking detects the direction of a participant's gaze, which in turn influences auditory cues, enriching their sensory experience as they navigate through the ambisonic soundscape.

Paine also extends creative agency in his interactive dance collaborations, where the dancer's movement is channelled into immersive vision and sound. In pieces like *Darker Edge of Night*, choreographer and performer Hellen Sky wears a muscle sensor on her forearm to activate Paine's interactive music composition. She hides certain control gestures and also performs more overt, gross gestures in a sophisticated multilayered choreography, Paine explains:

> I think a lot of the work that I did with Company in Space and with Hellen Sky and so on, and think about some of the works that we built where all of the music was, and in fact all of the music and all of the image world was coming off their body in real-time from biosensors that we had on her body, and so she was driving a work that was full 360° immersive projection and sound. She was basically creating off her body the entire environment that she inhabited and the audience inhabited, and so this is a really vast extension of the notion of performer agency from the more traditional, whether it's theatre in the round or proscenium. It doesn't really matter in terms of one's ability to transform the entire environment that's inhabited by the performer and the audience.

When working with skilled dancers, Paine was able to discover latent possibilities in the work that he never could have conceived, welcoming the opportunity to expand and share agency with movement experts during interdisciplinary projects.

Paine's early experiments with movement-sensing used aleatoric, or chance-based, musical methods, including controlling a synthesizer engine activated through floor pads and light beams to 'make the performance space a kind of living space where the work is being constructed in the moment rather than the form and the content having been fixed'. Installation pieces from the 1990s such as *Ghost in the Machine* question the relationships between behavioural patterns and the quality of an environment. Paine observed the response of public participants to an immersive audiovisual space and group decisions to behave collectively. The experience caused visitors to reflect on their own movement patterns, sometimes inspiring experimentation with cooperative ways of moving, like running around frantically in order to intensify the visual and auditory feedback of the installation through the magnitude and speed of their movements. Other times participants would negotiate with each other to remain still for extended periods:

> Even at that time, I was interested in how these questions of interaction allowed us to make broader sets of relationships between the audience, and also to some extent, to dissolve that sense of being an audience into being a maker in the moment who helped bring the work to life.

For Paine, an interest in the sets of potentialities that could be enacted has remained a pervasive theme in his performance and installation works. Depending on how audiences and performers engage with it, a work yields a range of outcomes. When participants are passive, the piece remains dormant. More active participation conjures exaggerated audiovisual feedback where, according to Paine, 'the thresholds of activity built different levels of engagement at different levels of intensity, and so how people enacted those spaces were dramatically different'.

In his performance works, Paine's gestural language is determined by the requirements of the piece rather than the instrument or system he's using. He adopts Trevor Wishart's (1996) dynamic morphology, a conceptual framework for understanding the evolution of the timbral, pitch and spatial aspects of sound over time, to frame his actions in relation to interactive and responsive systems. His movements reflect the morphology of the sound, such as the density of sound increasing with more vigorous physical movements. This fits with Paine's treatment of sound as a material that links directly to gestures in composition:

> Trying to get my composition students to think about sound as having mass and velocity and acceleration and density and surface and all these qualities so it becomes a very visceral material that you're sculpting, then that sense of being able to embody that was not hard at all, because you're no longer trying to embody some abstract notion of this has to go for eight beats, eight bars and then we do this and then this second inversion of that chord or whatever it happens to be. You're embodying the actual energetic state of the work itself, and so for me, any of the gestures relate to that.

Paine considers himself fortunate to have performed with a range of instruments that inspire different types of gestural interaction, including the Karlax – a sophisticated musical interface with ten continuous keys; eight pistons; motion sensors including an accelerometer, gyroscope and inertial sensors; and benders at both ends, similar in feel to a pitch bend wheel. The instrument captures and interprets the movement of the fingers, wrists, elbows, forearms and torso, enabling flexible control of several parameters. Moving it in the air generates inclination and acceleration data for sound control, while handholds allow the player to twist and rotate the two halves of the instrument. Numerous buttons provide tactile feedback and velocity sensitivity. Paine is particularly attracted to the clear and consistent somatic relationship that can be achieved with the instrument, offering the ability to layer music material and operate multiple controls simultaneously.

In a recent live work, *Music for Cymbals*, Paine performs the resonant frequencies of a virtual cymbal instrument using Roger Linn's Linnstrument MIDI performance controller. As opposed to the Karlax, which inspires a broader spectrum of energetic gestures, the Linnstrument was selected for its inherent tactility, invoking gently paced movements performed in an almost meditative manner:

> I'm playing these symbols that I've got resonators on, and I'm actually performing the resonant frequencies of the orchestra; symbols that are spread around the room, and so the Linnstrument is really great for that. The piece is very floating and very surreal in that kind of quality of sound and energy that fills the room, and so I can be very precise with the Linnstrument.

Having embraced a variety of commercial gestural instruments and gaming controllers, Paine resists the tendency in the NIME community to build a singular instrument or system to realize his musical ideas, feeling constrained by this personalized approach. Paine firstly considers the goals of a specific composition, before selecting an instrument. He does not compose a piece for a particular system, instead alternating between different interfaces like the Karlax; the Wacom Tablet, a control surface operated with a pen or stylus; or multiple wireless Nintendo Wii-Motes, gaming controllers with motion sensors that detect position, acceleration and angular data, which Paine uses to move in a nonlinear way through space and develop a morphological relationship between gesture and sound in response to the needs of a piece. The Karlax is favoured when a work requires 'very disjunctive kinds of gestures that have a really wide range of possible energy to them'. The Wacom tablet, like the Karlax, provides the ability 'to pinpoint the points in the timbre space and just jump between them and morph between them if that's what I want to do. ... So you could jump from space and have this very morphological relationship between gesture and sound.' Each interface provides a different way of physically engaging with it, promoting varied skills and forms of sound control, though repertoire for these instruments that don't produce sound will not possess the same constraints and uniformity offered by acoustic instruments (Paine 2015, 87).

Apart from the Karlax and Linnstrument, most of Paine's performance systems are customized, highlighting the absence of generalizable, universal gestural systems. Paine asks: 'Who goes down to the pub and just sees people with new interfaces playing gigs? And who knows people learning new interfaces?' Users of these specialized systems are more likely to appear at a NIME conference than on stage in a commercial venue. If they do appear in bars or concert venues, many audience members often do not fully comprehend how the performer's actions relate to the sounds being produced. Paine identifies the complexity involved in creating a ubiquitous instrument that can challenge and satisfy a broad range of musicians:

> The big roadblock is that in order to make anything that's actually truly interesting for anybody else to want to use and to make a really big body of interesting work, it has to be extraordinarily agile and it has to have a really large number of interrelationships that bring about instability and a certain level of controllable chaos in the instrument, and that's true in all good acoustic instruments.

To create an instrument that is flexible, imposing minimal limits on the performer, Paine (2015, 88) recommends a shift towards engineering solutions that prioritize the fit between the technology and performer's body, elevating 'idiomatic approaches to digital performance that focus on the performative potential and latent agency of the relationship between soma and techne'. Yet he finds that the structure of current systems is often restricted to shifting between fixed states:

> My concern is that in my view, even when we start to think about more complex interactive systems, we're still essentially talking about state machines, so we're talking about creating a certain subset of mappings and relationships that may under certain circumstances change state into another set of mappings. But within each of those sets of mappings, the parameter spaces are bounded, and the thing is that within an acoustic instrument that's not true in the same way. I mean sure, a flute is always going to sound like a flute, a violin always sounds like a violin, etc., but the parameter space of playing the flute can be bent and twisted into really unexpected and quite different spaces, so that suddenly the silver flute sounds more like a Shakahuchi and then it becomes more like a recorder, and then it doesn't seem to have these individual steps of pitch, but this sliding quality and so on.

To create a system that can navigate through timbre space, exploring creative potentialities, rather than switching from state to state, Paine has looked to acoustic instrument practice to discern the answer to the question, 'How do we bring all of those kinds of sensed qualities of expression into the domain of designing ways of performing computer-based music?'

The development of new electronic interfaces is hampered by the absence of a general model of musical control for such instruments, Paine (2009, 143) argues. To arrive at such a model, Paine researched the main aspects of successful performer–instrument interactions as part of the Thummer mapping project (ThuMP), asking, 'How do we get inside the computer and have nuanced control of things so we can play it perhaps with qualities that approach those that we find in and can develop in an acoustic instrument?' Interviews with musicians from orchestral to jazz styles investigated individual perceptions of instrumental practice, forming the basis for identifying key parameters of control musicians select when playing acoustic instruments, 'both in terms of the technique of playing an instrument and then in terms of music and musicality'. Paine feels that this overarching approach to control and expression has been missing from NIME research for a long time. ThuMP was the first large study examining the qualities of mapping in a practical context, resulting in a model for crafting a gestural language for controlling and generating live electronic laptop music that is informed by the musician's perspective rather than on an engineering or computer science approach. The resulting model identifies the physical controls of pressure, speed, angle and position as key variables influencing timbral elements like pitch, dynamics and amplitude, as well as linking to cognitive affordances (Gibson 1986) affecting playability. Paine then applied these control parameters drawn from the overlapping physical gestures that underpin traditional instrumental practice to creating design guidelines for further interface development in real-time live electronic music.

Paine consistently applies these design guidelines to developing works for a range of controllers, performance systems, VR applications and established instruments. His pieces form complex ecosystems in which performers and participants navigate immersive spaces, incorporating their somatic understandings and responses into collaborative musical outcomes. Paine's design ideas redirect focus away from technology and onto embodied relations with instrumental technology based on the phenomenological foundations of Merleau Ponty (1999) and Don Ihde (2013). Paine (2009, 142) contends that the experienced musician develops a proprioceptive relationship with an instrument, acquiring somatic maps that inform a partially unconscious perception of movement. It is this focus on the somatic that is the foundation of the next chapter, which reflects on personal engagement with touchless gestural systems that steer performer focus back to the body.

13

Intangible spaces

This chapter outlines my personal movement explorations and describes works that augment vocal and instrumental performance with body motion captured through remote sensing methods. Since 2008, I composed a series of solo and ensemble live electronic performances, exploring camera-based sensing of movement using a variety of motion sensors, from webcams to the Microsoft Kinect. Early experiments involved applying gestural control to processing the voice, then later exploring looping and temporal variations of instrumental and pre-recorded sources. These embryonic studies informed the design of the Telechord, a movement-controlled virtual instrument. Like a theremin, it is operated with touchless gestures; however, it also incorporates a freer movement vocabulary, polyphonic layering and visual feedback. It generates sound through physical modelling synthesis, filtering and augmenting harmonic overtones of the voice and building harmonies with virtual instruments. The Telechord is currently used primarily for digital signal processing, shaping the amplified, electronic imprint of the voice, as it is converted from analogue sound waves to binary data for manipulation. Looping and effects like equalization and delay transform the timbre and presence of the voice within space. Performances with the instrument integrate movement improvisation practices from dance and somatics, promoting flowing, continuous gestures akin to free dance.

Telechord

Technical set-up

The Telechord is an open-ended system for free movement exploration of sound synthesis and processing of acoustic and electronic sources. It is the foundation for developing new digital instruments and mappings while simultaneously exploring the potential of the moving body. There is no specific gestural vocabulary required of the performer, encouraging improvised and idiosyncratic movement approaches. There is no external object to hold, focus on, or relate to. In the absence of tangible feedback, a string instrument metaphor is implemented to guide interaction. In the first system version, four virtual strings are stretched across the apex joints of the skeleton. As the body moves, the length of the virtual resonating wires surrounding the body alter the pitch of four separate tones. The shifting relationships between selected joint positions produce harmonies that reflect the unique geometry of the human form. Chordal harmonies are produced directly with the body as virtual strings linking the limbs and head change in pitch when the body moves, affecting frequency in relation to the distances between selected joints.

The Telechord was modelled on geometric representations of the body's architecture and spatial dimensions dating back to Leonardo da Vinci's *Vitruvian Man* (ca. 1490) illustration and Pythagorean harmony. It links the body's proportions and ratios between the limbs, torso and surrounding space to activate harmonic intervals sonified by virtual physical models created in Modalys (Modalys n.d.), a physical model-based sound synthesis environment by the French Institute for Research and Coordination in Music/Acoustics (IRCAM). The instrument's mapping scheme is founded on familiar natural associations – in this case the geometric proportions of the moving human form and simulated vibrating bodies like plates, strings and membranes that can be matched with different exciters, such as bows and hammers. The name, Telechord, was developed to convey the communicative capacity of the instrument over a distance, signifying the transference of communicative intent between the performer and audience. It also references the relationship between the length of a vibrating string and its fundamental pitch, a discovery often attributed to Greek philosopher, Pythagoras. Through the monochord, a single-stringed instrument with a movable bridge, Pythagoras was able to study the harmonic relationships between vibrating strings of varying lengths (Caleon & Ramathan 2008, 450).

As there is nothing to grab hold of in this intangible, software-based system, I often find myself turning inwards, refining spatial and kinaesthetic awareness as I experiment with a range of postures and movement sequences. When improvising and performing with the virtual instrument, envisioning the imaginary strings offers a powerful mental image to guide my actions. The more I experiment with the system, the more it mirrors the patterns of my body. The sounds are a representation of how the assemblage of joints underpinning the physical architecture of the body interact. The vision and audio are the externalization of my primary instrument, which is my body.

There is a risk of falling into fixed movement habits. For this reason, I continually adapt mapping schemes and apply a variety of mapping ideas and dance and somatic techniques during improvisations to disrupt evolved gestural routines. During improvisation I always question myself – how do I feel during these movements? Inevitably, composing with movement and the body stirs up novel ideas and redirects stagnant energy. This feeling-based approach draws from the practice of dance and movement improvisation where kinaesthetic awareness is regarded as an essential part of the experiential body of knowledge (Blom & Chaplin 1988). Blom and Chaplin describe kinaesthetic awareness as a primary perception and self-awareness of the body in motion. The body's proprioceptive system judges 'spatial parameters, distances, sizes; monitors the positions of the parts of the body; and stores information about laterality, gravity, verticality, balance, tensions, movement dynamics' (Blom & Chaplin 1988, 18). They argue that awareness of movement grows through repetition and experience. Exploring movement patterns through improvisation refines sensitivity to emerging felt sensations, enhancing motion control, nuance and variability beyond established gestural patterns.

Focusing on the kinaesthetic sensations underlying movement, which function on a different level to the meanings and functions of gesture in a cultural context, offers an opportunity to reimagine habitual socially acquired bodily practices (Noland 2009). Within this kinaesthetic experience lies the capacity to transcend social conditioning, making way for new innovations in performance and cultural practice (Noland 2009, 2–3). Kinaesthetic experience, or awareness of one's own movement, can foster experimentation, dismantling bodily habits and challenging cultural meanings through conscious behaviour modification. This focused attention reveals the agency involved in overcoming an established set of bodily practices (Noland 2009). With every movement-based performance I continually overcome habits of movement that constrain my body. These include a routine of slumping when I play piano, sit at an office desk or sink into a couch, and an unconscious habit I gradually acquired of lifting my right shoulder when confronting technically challenging passages in piano performances. I was also prone to tightening my jaw during periods of stress. These detrimental habits resulted in unwanted tension stored in the body after the event, inhibiting flexibility and free movement. Through movement-centred performance techniques

I sought different ways of performing to attain more satisfaction and longevity through my practice.

The early design phase of the Telechord involved physically improvising with the system and roughly mapping continuous movement parameters including acceleration, position, velocity and distance between selected joints to virtual instruments in Modalys. Physical modelling synthesis building blocks of simulated material objects, including cylindrical pipes and strings, were paired with expressive 'empty-handed' gestures that are common in vocal performance and conducting (Cadoz & Wanderley 2000), often leading to unpredictable sounds, glissandos and uneven phrasing. The voice was also used as an exciter in conjunction with movement, introducing curved inflections into the sound by modulating their attack, frequency and timbre.

The Telechord uses four physically modelled objects virtually connected to selected body joints to produce sound by exciting each object proportionately to the joint motion at the point of connection. Motion thus has the effect of exciting the object while adjusting its virtual size (and therefore pitch). By altering the properties of the materials used to create each object, various tunings are formulated for each piece and edited in Max/MSP. Virtual instruments are designed by combining simulated plates, strings and membranes with different excitation methods. In Modalys software it is possible to create fantastical virtual objects, such as impossibly long cylindrical objects, measuring hundreds of kilometres, thus transcending the limitations of traditional acoustic instrument properties.

The rationale behind using physical modelling as the main synthesis method was to find a sound generation technique more closely aligned with playing acoustic instruments than other synthesis methods. Traditional sound synthesis techniques have incorporated oscillators, wavetables, filters, time envelope shapers and digital sampling of natural sounds. David M. Howard and Stuart Rimmel (2004) argue that physical models of musical instruments offer less abstract parameters that are more connected with musicians' experiences of playing traditional instruments. They recognize the need for electroacoustic musicians to maintain control over all elements of a sound and propose that physical modelling delivers a more intuitive method to achieve this, as the technique is based on the physical vibrating properties of objects found in everyday life, such as strings and membranes. The user is thus more likely to predict the result of a particular action compared to other synthesis techniques.

In *Real Sound Synthesis for Interactive Applications* Perry Cook (2002, xiii) writes, 'Our evolution and experience in the world has trained us to expect certain sonic responses from certain input gestures and parameters.' By incorporating these expectations into interaction design, it may be possible to establish a familiar grounding from which to develop coherent and believable mappings that are supported by understandings based on physical experience. However, when applying gesture to the control of physical models, the way in which sound is controlled can produce artificial effects that make its source difficult to detect. The challenge of creating a balance between the original sound's integrity and the gestural input thus became a primary consideration in the design of the Telechord.

The sound to movement mapping is deliberately simple and direct within the system. Virtual resonating strings linking the limbs and head change in pitch when the body moves, affecting frequency in relation to the distances between selected joints, as shown in Figure 17. Three basic types of physical actions produce sound on traditional acoustic instruments: blowing (also incorporating vocalizations), rubbing (including bowing and scraping) and striking (including plucking) (Cook, 2002). The Telechord supplements these actions with continuous and discrete improvised gestures applied to simulated physical materials. Positional data is obtained from the hands, feet and head. The distances between the hands to head and hands to feet are applied to modifying harmonic intervals of fourths, fifths and octaves. Acceleration calculated from movements control the excitation of each note, allowing additional expression from the force of motion. The underlying aim was to make technology transparent, relegating it to a supportive role while prioritizing the body. Unlike particular cues that are recognized by gesture-recognition systems, this kind of system, which analyses continuous flowing gestures, does not require radical alterations to movement. Magnitude, energy and the pace of movements are coupled directly to the intensity of sounds, highlighting physicality and the centrality of the body.

The visual feedback displays the virtual physical models and strings depicting connections between key body joints. This geometric simplification of the human form is designed to reflect 'geometry that gives us a tangible image of space' (Newlove & Dalby 2004, 23). The first version of the visualization is a customized open-source program called Smokescreen. The amplitude of the instrument's audio output and the positional data of the captured joints are depicted as a two-dimensional (2D) smoke simulation, showing the amount of

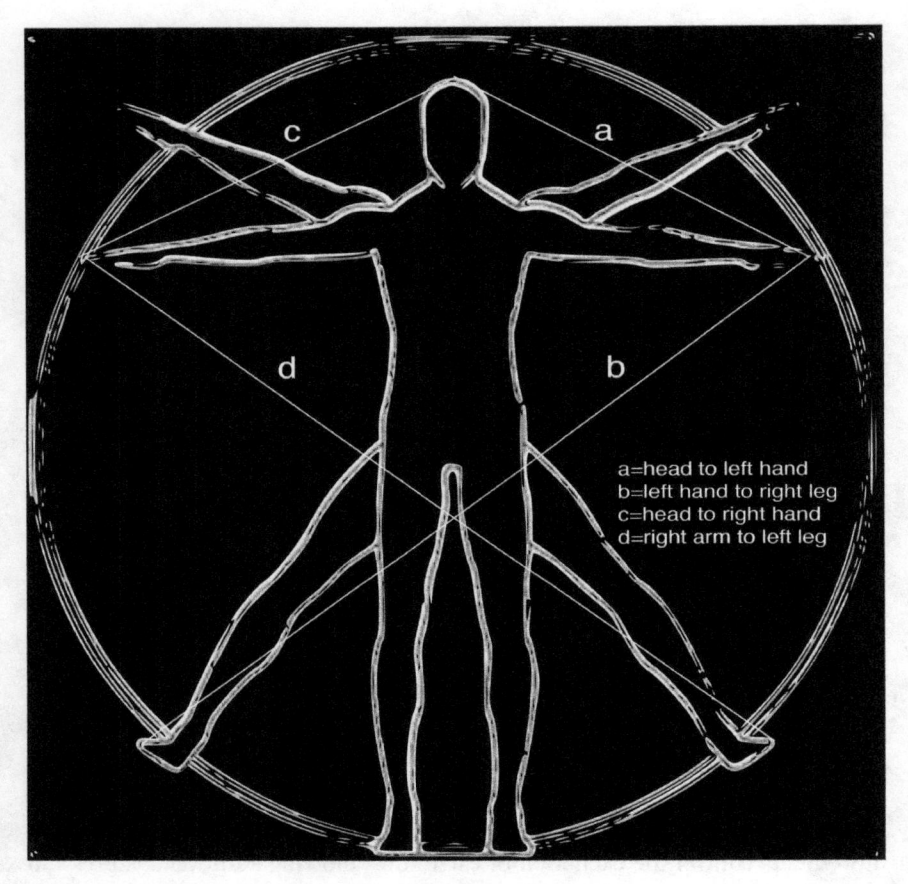

a=head to left hand
b=left hand to right leg
c=head to right hand
d=right arm to left leg

Figure 17 Virtual string positions of the Telechord

energy being injected into the system. Smoke particles are injected into a 2D fluid simulation, where the fluid is disturbed by their motion. Tendrils of smoke emerge, responding to the voice and instrument amplitude, movement velocity and acceleration data. Over time, the fluid disturbance decays to zero and the smoke particles diminish in brightness. RGB coloured lines were later added, connecting the head, shoulders, legs and arms. The lines are subject to a slight time delay, creating an echo effect, seen in Figures 18 and 19, which visualize the past 200 positions of the body. The video output maps the traces and echoes of movement, signifying the paths where movement has once been, reminiscent of Eadweard Muybridge's motion photography depicting patterns in human and animal locomotion. Movement traces a path that is constantly altered by

Figure 18 Visual feedback of the Telechord. Photo credit: Dean Lewins

previous actions and amplitudes. The imagery is intended to echo the nature of musical memory while indicating to the performer the positions they have recently visited.

The visual feedback nurtures new ideas by reflecting the nuances and energy inherent in performer movement. It also provides another layer of connection with the audience. Buxton (2007, 136) emphasizes the importance of sketching ideas in a range of formats to draw out key aspects of interactive design that 'capture the essence of design concepts around transition, dynamics, feel, phrasing, and all the other unique attributes of interactive systems'. Visualizing movement is also illustrative, helping to accentuate the geometric forms underpinning the interaction design of the Telechord while providing insight into how the moving

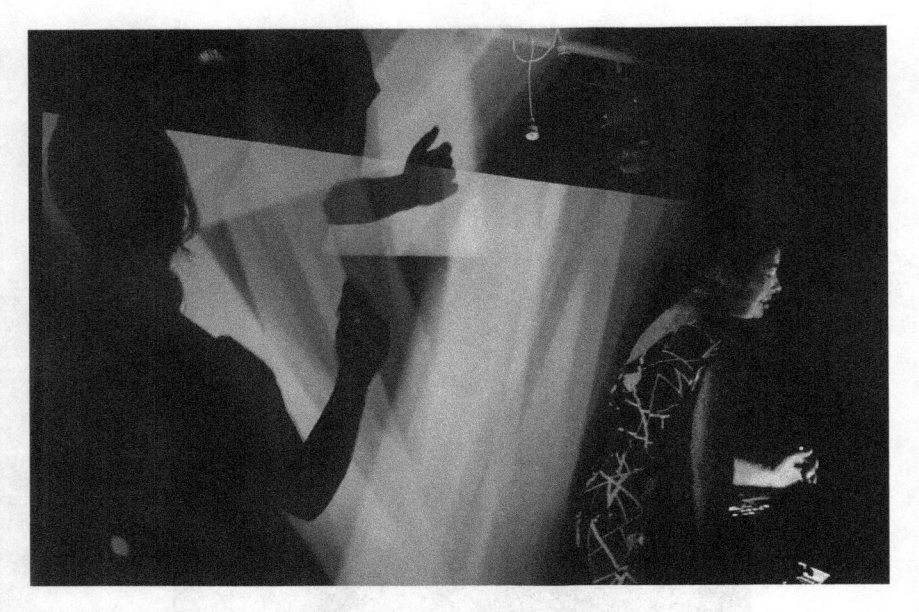

Figure 19 Visual feedback at *Intangible Spaces* (2018) performance, Vivid Sydney Festival. Photo credit: Rhiannon Hopley

body is being interpreted by the system. Visualization as a tool aids the design and composition process, making movement more accessible in a manner akin to sketching ideas on paper early in a system's conception (Hansen 2011, 252). Interacting with the visual feedback during improvisation magnifies movement qualities usually lost in temporal motion. Similar to ballet dancer Marion Cavaillé's experience of improvising with an abstract visualization of mass spring physical models in *Chiselling Bodies*, I came to build a relationship with the projected visual model of a virtual string instrument framed by the body's skeletal structure, almost as if it were a separate performer, through ongoing improvisations with it (Fdili Alaoui, Henry & Jacquemin 2014, 177). By using projected visual imagery in association with embodied mapping, the Telechord is designed to gain form in the player's experience, assuming part of an acquired body schema.

The integration of embodied principles and metaphors in mapping performer movements to sound processes is growing in momentum (Antle, Corness & Droumeva 2009; Wilkie, Holland & Mulholland 2010; Roddy & Furlong 2013), demonstrating the benefits of borrowing from established links and embodied understandings of physical phenomena to increase the chances of developing a gestural system that makes sense and feels natural to the performer. Antle, Corness & Droumeva (2009) refer to links between abstract concepts and our physical way of relating to the world as embodied metaphors. When designing mappings, they suggest that tacit knowledge of a physical source domain can inform a conceptual metaphor to aid users in interacting with a more abstract conceptual target domain (Antle, Corness & Droumeva 2009, 240). Evidence from an exploratory study comparing interaction with and without embodied metaphors reveals that the embodied metaphor interaction model is more intuitive for users (Antle, Corness & Droumeva 2009). Informed primarily by the theories of George Lakoff and Mark Johnson (1980, 1999), metaphor-based sonic interaction designs draw on sonic expectations formed from prior bodily engagement with everyday auditory phenomena. Lakoff and Johnson argue that individuals interpret abstract concepts through the lens of sensorimotor experience, viewing the 'metaphor' as a way of establishing meaning by relating one concept from a 'source domain' to another in a 'target domain.'

The related concept of image schema refers to patterns formed from internal representations of the body and its movements. These patterns derive from past physical experiences. Johnson presents the concept that physical experiences contribute to the formulation of structures and patterns he calls image schemas in *The Body in the Mind* (1987). These basic structures of sensorimotor experience 'define the contours of our world and make it possible for us to make sense of, reason about, and act reliably within this world' (Johnson 2007, 136).

In the Telechord the internal imagery of the musician is reinforced by layered visual feedback that simultaneously depicts the strings connecting active joints and a 2D particle system that traces paths of performer movement. Within the dance field, Shannon Cuykendall, Thecla Schiphorst and Jim Bizzocchi (2014) advocate the benefits of using images in performances to also engage the audience more kinaesthetically, finding that iconic images can emphasize the subtleties of movement, potentially engaging a broader audience. This approach has also proven effective for Sergi Jordà (2003) when displaying visual feedback of the Faust Music Online (FMOL) software with a projector for the audience, in exposing the underlying musical processes to the audience without the need for program notes or extensive explanations.

Improvising with a bank of virtual instruments accumulated over several years involves calibrating how the range of motion will affect the envelope and evolving intensity of the sound, while refining pitch consistency, shaping glissandos and taming dissonances. Public performances involve embracing technical glitches and randomness in varyingly lit and amplified settings. Minute muscular actions often lead to small sonic oscillations and jittery visualizations. Focusing on kinaesthetic feedback in relation to micro-movements feeds 'highly individualised imagery strategies for generating movement and exploring the feedback loop' (Francois & Schiphorst 2017, 5172). Consistent movement patterns emerge, demonstrating the discovery of a unique movement language comprised primarily of cyclical and traverse arm movements and torso twists that capitalize on the features of the Telechord design.

Performing intangible spaces

The din of the rising and scattering layered virtual physical object plate models accumulates against the advancing drums and bass. At times I am lost in the frequencies. I focus my attention most effectively in the minimalistic solo sections. Slowly fluctuating drones develop in response to my primary emotions and energy levels. I let them reduce to a croak. Guttural sighs and groans release energy. The filtering effects of wind-like and simulated breath noises amplify the emotional intensity further. I slice through the air, my arms tiring easily, generating short synthetic phrases that ring out into a virtual cavernous space. I mutter, sigh, guffaw and speak with invisible entities. The voice sometimes becomes tangled in the metallic texture of multiple scraping physical plate models. Throughout the forty-minute set it is challenging to accurately perform regular pitch transitions, and to maintain consistency and repeatability. I need to tune into feel to recognize how precise I can be. I wind down to complete silence then build up the din again.

The ensemble performance work *Intangible Spaces* debuted at Intangible Instruments, an event exploring movement-based works and systems with Donna Hewitt and Julian Knowles at the 2018 Vivid Sydney Festival of Light, Music and Ideas, depicted in Figures 19 and 20. It was later presented as a solo performance at the Sound and Music Computing Conference (SMC) in Cyprus. In the latter version I relinquish fixed compositional structures to freely improvise, letting non-verbal utterances rise from deep in my chest cavity and throat. I abandon pre-composed lyrics, using longer, sustained phrases to articulate pure, unprocessed emotions. The visual is more exact – it shows thinner lines like steel wires that display shifting relationships between key body

Figure 20 *Intangible Spaces* (2018) ensemble performance, Vivid Sydney Festival. Photo credit: Rhiannon Hopley

joints. However, I cannot see the visual feedback as the imagery is projected onto a screen behind me. Usually I gain a faint impression of the feedback from a semi-transparent scrim positioned at the front of the stage, still being able to see and interact with the audience. Without it, focus moves to internal bodily sensations instead. I extend and fold my limbs in a series of stretched, contorted and crouching poses. I complement this improvised movement with a gestural vocabulary characterized by simulated bowing, striking and plucking gestures. Introducing the constraint of monophonic control, which mimics the interaction design of the cello, the left hand controls frequency in vertical space while the right hand bows or plucks a single virtual string horizontally. The levels run hot in response to the dynamic input of the bellowing voice feeding into a plate model.

Youbeme

The challenge of fixing the temporal world of improvised gesture-controlled musical performance in recorded form has involved compiling several years of recorded materials for a live album, crafting arrangements aimed at preserving the spontaneity of the vocal and movement-based improvisations I have collected and archived. As Sonami previously mentioned, it is difficult to capture the process of generating and transforming sounds through such a temporal and transient art form as movement. Yet I experienced the benefit of recording improvisations and performances for later analysis, as the multisensory experience can become overwhelming and passes in what seems like a flash of light during a public performance where only snatches of feeling and audience receptiveness can be remembered. Other occurrences can be immediately imprinted on the body, such as a glance or poignant comment from an audience member.

The album, *Youbeme*, includes substantial portions of the 2019 live work for voice, drums, electric bass and Telechord, which was first performed at the Electrosocial Exhibition presented and curated by independent art organization, Electrofringe. The piece explores the notion of kinaesthetic empathy by blurring audience and performer distinctions, inviting them to sample the first-person felt perceptions of the performer. The audience can capture and log footage of the event with mobile phones, electing to share their viewpoint on an online media server that merges a range of viewpoints of the event on a timeline, including the musicians' perspective. Performers film the event with wearable camera glasses. On a cramped stage, I am joined by Robbie Mudrazija on drums and Meeghan Oliver on bass, improvising bass heavy rhythms with the uneven phrasing of the Telechord and voice undulating in between. Vocal fragments, heavy sampled percussion, and distorted and synthesized bass meet gestural improvisations with effects augmentation. The heat of the battery-operated glasses and projectors beaming the strings onto my body, together with constant movement, bring my core temperature up as the show progresses. The lyrics are compiled from a random assemblage of free association phrases overheard from conversations leading up to the event. The instrumental visuals are complemented by the saturated rear projections of video artist J D Young, seen in Figure 21.

Youbeme is an embryonic version of my vision to develop augmented reality works and participatory experiences with wearable technology and mobile phones, part-time art making devices that most individuals carry everywhere and are capable of capturing high-quality audio and video or gallery ready prints. Like in Paine's *Future Perfect* (2018), phones become as ubiquitous as bodies, inviting participation from the audience.

Figure 21 *Youbeme* performance at Electrosocial exhibition with rear video projections by J. D. Young. Photo credit: J. D. Young

Magnetic springs

Like other recent works, *Magnetic Springs* looks at ways kinaesthetic feedback and new ways of moving can be developed. The piece examines the interaction between four different types of feedback: kinaesthetic, visual, auditory and tactile. After many years of non-tactile, intangible control, I wanted to incorporate additional physical resistance and feedback to the Telechord, for myself as much as for the audience, who often wondered about my mysterious, choreographed spatial gestures on stage. I wanted a central point of focus and an object to orient my actions towards. Building on the string metaphor of the Telechord, the strings materialized into steel springs, acting as exciters for various models of vibrating plate, pipe and membrane objects. The common child's slinky toy, with its classic solenoid structure, was appropriated to evoke a playful approach. The sounds were mapped to four speakers to create a quadraphonic mix that channelled four string sounds to opposing corners of the room.

During the *Magnetic Springs* performance at the closing party of the 2020 Tangible, Embedded and Embodied Interaction (TEI) conference at Sydney University, I explored haptic feedback through the gyrations of the steel springs. I mounted jumbo- and regular-sized slinkies to a resonating wooden box equipped with contact microphones. Two contact microphones transduced the acoustical vibrations of springs of varying lengths and diameters to electrical signals that formed an input or excitation for Telechord sounds. Yanking and stretching them, I felt the tension, resistance and rebound of real 'strings'. I later learned of Pamela Z's percussive slinky experiments in the mid-1990s that explored similar principles by repurposing commonplace objects in live electronic music. My version became playful and slightly ridiculous, prompting striking and swinging movements that were far more expansive and exaggerated in performance than during rehearsals. I pulled, plucked and twisted the springs with repeated circular motions to create overlapping sustained sonic textures inheriting the metallic qualities of the steel coils. I prowled the full stage space during the eleven-minute performance, dragging the resonating wooden box equipped with the springs about.

Constructing increasingly dense polyrhythms, the piece establishes a bridge between the tangible and the intangible. Wishing to progress from metaphoric string instrument gestures, I instead engaged with the physical properties and movement tendencies of the object. The instrument in this case becomes an extension of the body. Using the two springs to excite the physical modelling synthesis sounds crafted in Modalys, I was able to gain more immediate and direct control of the attack and decay of sounds. Shaking and plucking an actual spring, compared to simulating physical gestures through spatial movement offered greater scope for musical expression and the ability to articulate the envelope of sounds in more detail. The springs inspired expansive upper body motion and improvised lunges with the lower part of the body.

I feel the internal energy levels rising before performance. Adrenaline starts pumping. I find the technology, and what could possibly go wrong, distracting. I always feel like this when performing with a new or radically modified system for the first time. My senses are heightened. I feel hot and cold simultaneously, hearing sound more intensely. I aim to go into myself and tune out the external world. My body, in concert with the equipment, unites the internal cognitive and fleshy parts. My breath and voice, partnering with body motion, discover a new language that meets the resistance and recoil of the springs.

Reflection

I reflect on what Sonami mentioned about the pressure to be seen as an original technical innovator in this field. This view ignores novel applications and unique performance techniques developed to gain the most from performance systems. It glosses over the meaning and messages behind the work itself, or the way it explores novel modes of performance, which can take decades to realize. For artists using the same systems for years, this persistence can sometimes be seen as a weakness, interpreted as an inability to conceive of new ideas, Sonami observed. There is little recognition for refining and mastering skills with one personalized instrument. Yet if musicians keep designing new and improved systems, playing skills inevitably suffer. Like Sonami, I will move beyond the Telechord once it becomes predictable and when I have learned everything I can from it. Cannon also grew with her meta-woodwind instruments, playing ornate digital bassoons that blend the expressive characteristics of both the bassoon's double reed and the amplified inner breath-like sounds that are usually hidden. Like virtuosic jazz player George Lewis, who uses predictive software to improvise with, Cannon is a virtuoso of the Serpentine Bassoon and Contra Monster as a result of investing in playing them over the long-term.

The original inspiration for the Telechord was based on somatic and improvised dance practices that build movement sensitivity and awareness. I have been playing it for approximately eight years. Wilson-Bokowiec sees gestural practice as an art form that is still emerging. It is never fixed or complete. Systems and instruments are always in flux. Musicians are constantly developing physical skill-sets and inclinations. The live album featuring the Telechord is a record of performances and rehearsals, each inspiring new design iterations, the last seeping into tactile territory. All of these live works take time to program, test, prototype and rehearse. Recording these efforts captures traces of time, an artefact or imprint of fleeting sound-movement improvisations.

Like the ephemeral nature of gestural improvisation, the supporting code and third-party plugins are also temporary. Since I started using the Kinect in 2012, I've adapted multiple middleware applications, from Ryan Challinor's Synapse to Jon Bellona's SimpleKinect. Repeated operating system and software updates lead to challenges in freezing development and archiving the artefacts of these intangible systems. Audio recording at least documents the movement patterns emerging in response to these rapidly changing instruments that never yield the same musical outcome twice.

When I move I often assume the shape of melodic lines with the arms or through torso or hip contortions. Glissandos and rapid pitch ascensions and descensions dance. I feel the stuck parts shifting within my body. When I match that movement with singing, my mood is lifted by the intake, expulsion and increased pace of breath, often leading to a light-headed, giddy feeling. I notice the inbuilt tensions arising and move to counter them, exercising body-based resistance. I work with past inner stresses that have accumulated from my predominantly sedentary lifestyle. After observing my voice in everyday conversation and its downward pitched phrases and lack of projection on recorded conference calls, I consciously modify recurring speech habits by intentionally modulating the inflections and energy behind them. Those alterations then have an effect on my thoughts and relationship with my body, elevating and expanding them. When I inject more energy into my voice, it appears to have a broader effect on my being, lifting my overall mood. This deliberate process reverses decades of socialization and experience that has moulded my voice as it is now, transforming it into a self-made invention that in turn, interacts with and is emblematic of my body.

Occasionally I will coax myself through yoga sequences delivered online. Though I can't summon the motivation to sustain a daily or even regular exercise practice, I attempt to engage my body in physical pursuits through activities that don't feel like exercise, like gardening or walking on the beach. When motivation wanes, I sit moulded to the couch, concave, writing and web surfing, wondering about the effect of this artificial slump and slowed down activity on my thoughts and health. It must slow the heart and blood flow through the circulatory system, and particularly to the brain. The curved inward posture must affect my breathing and intake of oxygen. Contrasting this static state with singing, a practice that signals deeper bodily engagement, I realise that I become more attuned to the depth and pace of my inhalations and exhalations when vocalising. I need the intensity of performance to extend my vocal cords and realign my posture. More than any other instrument, the voice encourages an upright posture and an awareness of physical sensations to project sound further, reaffirming Gilman's (2019, 76) observation: 'How we move, stand, and sit is part of what makes our voices sound and be able to do all the wonderful things we can do vocally.' Like singing, dancing spontaneously or a casual walk in nature often encourage new ideas for me. This newfound inspiration is perhaps due to the variation in movement or the increase of blood flow throughout the body. These activities heighten awareness of movement, matched by the expanding and receding rhythms of breathing.

Free movement: Free music

For my current set of works, compositional ideas spring mainly from the irregular rhythms of everyday life – derived either from improvised body movements, sounds heard in nature, or from random human-made or machine-made noises like random analogue synthesizer oscillators that never repeat the same patterns twice and morph subtly over time. These ideas are prominent in Grainger's notion of free music, prompted by an observation of the undulating movement of the sea (Linz 2003) and manifested in a work for four theremins, *Free Music No. 1* (1936). The parallels between gestural music and this piece include curved and irregular pitch transitions. In developing the Free Music Machine with Burnett Cross, Grainger aimed to hear the music of his imagination, first conceived in his boyhood, with no restrictions of quarter tones or half tones. Grainger aimed to produce complex uneven rhythms that a human performer could never faithfully reproduce, allowing the composer to communicate with listeners directly rather than channel works through musicians. This is the point at which movement-based musical systems deviate from Grainger's vision. Usually gestural systems incorporate software that *reflects* the innate human variations present in a performance and the rhythms of human actions. The Telechord, for example, captures inherent syncopations in individual body movements. It does not replicate machine rhythms. I want it to embody human and natural rhythms. Rhythm to me is a form of directed energy. Internal rhythms of breathing and circulation underpin our existence, such as the natural rhythm of wakefulness and sleep, which establishes a constant ebb and flow throughout our lives (Sheets-Johnstone 2010). These inbuilt rhythms converge with broader social rhythms, regulating patterns of human activity while providing vital clues to understanding the idiosyncratic nature of movement.

Engaging in free movement has enabled me to learn more about my own movement patterns and find ways of transcending them, channelling the immediacy of embodied experience in music. I have come across fortunate coincidences, embracing the unpredictability of inventions emerging from physical expression. Unintended glitches often produce the most interesting results, though are not always repeatable, causing frustrations in live work. Both voice and unregulated spatial instruments like the theremin lend themselves to what Grainger referred to as free music. The gliding tones of his free music machines correspond with Trevor Wishart's (1996) non-lattice classification based on the fluidity of the voice compared to instruments of fixed pitch like piano keyboards. This move away from individual note-based performance is seen by Tellef Kvifte (2011) as a change that occurred between the nineteenth and twentieth centuries, placing a new focus on timbre. Similarly, the space surrounding the body cannot be dissected into grid-like sections. It is constantly morphing through time and reflects unique individual perspectives. The movement abilities, patterns and preferences of the performer determine how they move through space.

Designing and performing with gestural systems has uncovered a new type of hybrid practice – sound-movement improvisation. Similar to movement improvisation in contemporary dance practice, this has become a personal method of ideation, warming up the body and refining and calibrating it to the interface. This is often accompanied by verbal ramblings, vocal utterances and guttural groans and sighs. These sounds are released from deep within the core of my being. During improvisations, I notice and observe how movements and vocal sounds interact and resonate sympathetically with each other. Just like speech, my gestures help to express and draw out half-formed, abstract concepts.

Unlike dancers, however, who are often dancing to musical or sonic accompaniment, I generate and compose with sound and choreograph movement in real time. The practice has opened up awareness of my body's internal resistances, strengths and weaknesses. I can gently ease out of resistance and limitation more effectively when I move than when I sit still. When I have a sore back or shoulder from excessive exertion, resting is often more painful than slow, gentle movement. Movement also evokes novel ideas in design in a way that sitting watching the computer screen does not. This may explain the increasing popularity of embodied and vocal sketching, which invests the entire dynamically moving body in the sonic interaction design process. Vocal sketching involves the voice and body in generating correspondences between action and sonic outcome (Frid, Elblaus & Bresin 2019, 182), while embodied sketching enables musicians to explore potential relationships between movement and music through direct physical engagement with an interface.

Sound-movement improvisation is a related prototyping, rehearsal and performance method that offers an opportunity to determine the scale of my movements in relation to sound and discern links between specific movement qualities, whether they be swift, sweeping, expansive, minute or consciously slow, and specific sonic parameters. Dancers in the interactive realm compose with their bodies, usually in collaboration with composers who assemble and supply precomposed material or sound-generating systems. As movement experts, they can usually yield the most nuance from these motion-based performance systems with their finely honed techniques for exploring and varying movement and working expertly with the weight and shape shifting forces of the body, particularly in collaborative contexts. Without the formal movement training of a trained dancer, sound-movement improvisation offers me a pathway to develop these essential skills for gestural performance.

Since first adopting improvisation as a central focal point of my performance and design practice, I have become more attuned to the qualities of movement – recognizing the lightness or heaviness of a particular turn or sequence. I have explored the characteristics of symmetry and balance. I have expanded my physical endurance and movement imagination in relation to sound. I have experimented with connections between specific gestures and sounds to establish new sound-movement mappings. I have developed more variation and willingness to explore movement. Unlike a trained dancer, I have developed these skills independently, primarily as a solo performer. In the future, I aim to work more with dancers, choreographers and somatic practitioners who may provide further tools for enhancing this ongoing exploration.

I feel aligned to Isadora Duncan's approach to natural forms of dance, which laid the groundwork for modern dance. I want to move in a way that the body naturally wants to go. I have no 'overlearned' movement vocabularies. When I explore and improvise, I continually discover new ways of moving and acknowledging gestural habits. Moving while making music takes my body even further into inhabiting the music. When combined with the internal mechanisms of the voice, the hybrid practice of vocalizing while gesticulating to create sound can be incredibly powerful. Although initially resistant and challenged as a classically trained vocalist, Shannon Holmes (2016, 198–9) eventually found the value in free improvisation. Improvisation, in association with a focus on the physical sensations of movement, sound and text, became profoundly transformative within her practice:

The use of improvisation to evoke images became a starting point in the process to mine an autobiographical narrative, and my body indeed became a co-performative agent in the process.

Free movement or improvisation also acts as a springboard for the invention of novel choreographies in contemporary dance. From Merce Cunningham to Martha Graham and their predecessor, Duncan, improvisation becomes a method for performers to connect with their inner natures and the natural world, as a method for magnifying and subverting habit. While Duncan took her cues from nature, Graham and Cunningham embraced angular lines and trajectories, deliberately disrupting the pathways their bodies would usually go.

The main benefits of improvisation are thus overcoming habit, generating novel material, and enhancing sensory awareness of the body in motion. Sheets-Johnstone explores the potential of improvisation to promote kinaesthetic awareness in her book, *The Primacy of Movement* (1999), highlighting the felt qualities of human motion, which are perceived through the kinaesthetic sense. She argues that in becoming attuned to these qualities by performing free variations of habitual movements, a greater awareness of personal movement patterns can be gained (Sheets-Johnstone 1999, 143). Attention to feelings associated with physical action, 'which include a bodily felt sense of the direction of our movement, its speed, its range, its tension' (Sheets- Johnston 1999, 56) develops with experience, advancing understandings of how bodies move in the world. Sheets-Johnstone (2010, 217) emphasizes the overall neglect of the experiential aspects of movement in research on embodiment and enaction, blocking a deeper understanding of individual physical expression.

Improvisation can also be associated with play – low-stakes creative experimentation with no fixed outcome that is often fun and frees the individual from paralysing inhibitions, as composer and performer, John Ferguson writes:

> 'Play' is always at the forefront of my thinking and doing; these are not separate activities, because any attempt to design, build, or reconfigure, is always mediated by play, and this helps retain the immediacy of music-making. For example, when intervening in electronic circuitry through hardware modification, musical expressivity can potentially be far greater than what might be expected if the same circuitry is approached in the manner that its designers intended. (Ferguson 2013, 137)

In Cannon and Favilla's article, 'Investment of Play' (2012), they demonstrate a substantial investment involving hundreds of cumulative hours in playing their instruments. The notion of play as opposed to practice, which implies rote learning and repetitive reinforcement of instrumental technique, encourages exploration and experimentation. The artist leads the creative and design agenda surrounding novel gestural systems, following no preset program to limit their creativity.

Musical and movement improvisation in my own practice offer avenues for attuning the sensory body to internal and external stimuli and discovering meaningful mappings between movement and sound. Bearing many similarities to embodied and vocal sketching, improvisation is a key method for linking performance and design activities. To this end, the Telechord was developed as an open-ended system catering for movement and sonic improvisation that spans across performances and recorded media. Drawing on Paine's techno-somatic design approach, it aims to achieve an intuitive fit between body and technology. An unexpected side effect of the experiential prototyping process undertaken to achieve this goal was the development of a heightened sense of kinaesthetic awareness and sensitivity to my own spatial potential and movement patterns. Like Chris Salter (2012, 182) and collaborators in the creation of the *Just Noticeable Difference (JND)* movement-controlled installation, direct bodily engagement played a vital role in the programming and refinement of the system. These tasks could not be undertaken from a distance given the physical nature of the interaction. Adopting a similar design method assisted me to formulate design goals and guidelines in accordance with my artistic and pragmatic performance aims.

Like many of the other artist-designed gestural systems explored in Part Two, the Telechord is highly personalized to match individual movement patterns and compositional strategies that evolve over time. In the third part of the book, I discuss the potential of long-term experience with gestural systems to influence a musicians' perceived identity and relationship with their body. The implications of using gestural systems for musicians in broader professional practice is also examined.

Part Three

Synergy and transformation

The final part of the book examines the broader cultural impacts and transformative effect of gestural systems on musical performance techniques, identity and idea generation. The unique experience of playing a gestural instrument, which only a small percentage of musicians experience, can prompt unique ways of experiencing music physically. Emerging and established motion capture technologies have altered the way musicians create and relate to their bodies. These experiences help performers to address and transcend their usual movement habits and, in turn, their ways of thinking. The following discussion refers to recent theories of neuroplasticity, which reflect on the brain's capacity to change and adapt in response to individual experiences, practices, sights, sounds and other sensory information, creating new connections between neurons. This research provides insight into how embodied engagement with novel interfaces can expand vocal agency and motor skills while also shaping identity.

The potential of gestural systems to reinforce awareness of the body, extending established notions of instrumental and physical control, is evident in the first-person accounts of artists presented in Part Two, showing the long-lasting effects of performing with gestural musical instruments on movement habits and understandings. Engaging with systems that reflect physiological data back to the user can promote heightened self-agency and awareness (Nunez-Pacheco & Loke 2014), feeding new creative discoveries, altering identities, and extending movement vocabularies and abilities. Movement and voice form fundamental components of identity and individual differentiation. In tandem, they broadcast an individual's health, mood, abilities and energy. A willingness to explore both modalities through experimentation with gestural systems can strengthen sensitivity and the expressive capacity of musicians in live electronic performance.

The structure and stamina of human bodies are the product of repeated actions. Individuals can also be transformed by indulging in varied movement patterns. Gestural performance is an expanding area that allows musicians to engage and inhabit the body in novel ways of their own making, offering the potential to reinvent individual physical aptitude and form. Participation in movement-based electronic music requires refined kinaesthetic awareness to improve movement expertize, reflected through enhanced movement variation, economy and efficiency. For Shusterman (2009), body consciousness and awareness precede transformation and empowerment. Shusterman's call to abandon habitual actions that no longer serve the body can be replaced by greater openness to movement variation. The harmful physical routines many people perpetuate, according to Shusterman, often involve habitual slumping, which over time yields a stooping posture and limited flexibility. Becoming attuned to the body's inner resistances provides opportunities to counteract inbuilt tensions and achieve heightened movement performance and endurance.

Limited movement abilities result in limited movement repertoire (Höök et al. 2016, 3132). To overcome this restriction, it is necessary to broaden participation in a range of movement activities and engage in critical self-reflection throughout. Adopting principles from bodywork practices that emphasize felt awareness can inform this process, Höök argues: 'The concept of *self-agency* in somatics practice is the key. Self-agency is the result of the reflective practice of self-observation coupled with intention' (Höök et al. 2018, 6). As Moshe Feldenkrais (1972), the founder of the Feldenkrais method, posits, an individual's self-image shapes their thoughts and movements. Feldenkrais regards the systematic transformation of self-image, comprising the four components of movement, sensation, feeling and thought, as more effective than a piecemeal approach that focuses on changing the body through one action at a time:

> Improving the general dynamics of the image becomes the equivalent of tuning the piano itself, as it is much easier to play correctly on an instrument that is in tune than on one that is not. (Feldenkrais 1972, 24)

A key aspect of Feldenkrais method is to conduct a holistic analysis of habitual movement patterns. If any actions produce pain or discomfort, the overall movement approach is analysed and altered to promote ease. This conscious reflection is primarily achieved by focusing on and interpreting the felt sensations of the body.

There are a number of ways performers can tune into the felt dimension beyond pursuing somatic practices or undertaking any activity while focusing on how the body is feeling. Another method for honing awareness is conscious listening. As previously discussed, composer Pauline Oliveros promoted a type of listening that engaged the whole body. In a similar vein, Hildergard Westercamp, a Canadian composer and acoustic ecologist, explores the potential of the ears and body to relate to sounds of the lived environment. Both artists found listening through the body permitted transcendence of societal ideals:

> by allowing our bodies to connect to the whole sounding world, we can bypass institutional lines and frames, consequently demolishing the boundaries between the outside and the inside, between a man and a woman, between nature and culture. (Kazlauskaite 2020, 349)

In Westercamp's work *Breathing Room* (1990), she observes the connection between repetitive breathing, speech and gestures on her relationship to the outer environment and interpretation of her surrounds. She discovers how closely breathing is related to heartbeat and how it does not exist in isolation. Westercamp's inhalations and exhalations create sounds in her environment, becoming a social phenomenon that melds with the soundscape of her overall world (Kazlauskaite 2020, 351). Breathing and listening become transformative through heightened awareness of the fundamental processes. Attentiveness to breath in meditation practice is what yields transformation of thought patterns and mood, not the breath itself. Breath is normally an automatic activity. It is the attunement to sensations and bodily patterns underlying these involuntary processes that shift individuals into a different state of being.

Oliveros (2005) also endorsed an open-ended, body-led and improvisational approach to the experience of conscious listening, which she termed Deep Listening. Integral to this approach are inclusivity and self-reflection. Insights are unearthed through breathing exercises, vocalizations and dreamwork to access intuition. Rather than thinking through and organizing sound in musical terms, as advocated by John Cage, sound can be processed intentionally through the body to unveil inner truths and connections between listeners, enhancing social inclusion (Kazlauskaite 2020, 349). For Oliveros, this embodied consciousness expands listening to a communal experience that draws on internal body knowledge, disturbing the usual channels of power in traditional approaches and institutional structures that dictate a purely intellectual approach to sound – as a medium that can be tamed and controlled (Kazlauskaite 2020,

351–2). By embracing varied types of listening, including neglected and forgotten techniques, established forms of auditory thinking can give way to individual-led listening practices: 'New fields of thought can be opened and the individual may be expanded' (Oliveros 2005, xxv). This consciousness fosters an inclusivity that boosts creativity by welcoming the experiential insights of diverse participants.

These transformational activities allow individuals to create a new world of sound and sonic imagination, rather than frequenting previously trodden paths of a predefined masculine world in order to 'surpass its gendered and dividing structures' (Kazlauskaite 2020, 355). What if, in enlarging dominant perceptions of sound, it becomes possible to liberate the body from fixed conceptions? Feminist writer Ursula Le Guin advocates for an experiential and feeling-based approach to storytelling that echoes Oliveros's deep listening philosophy, arguing, 'When we women offer our experience as truth, as human truth, all the maps change. There are new mountains' (Le Guin 1989, 160). Similarly, embodied awareness borne from tuning into the feelings underlying everyday movement can disrupt the habitual actions of an increasingly narrow life, suggesting new ways of exploring the body's expressive potential. This exploration leads ultimately to shifts in identity and agency, societal roles, and self-perception. Sharing these altered narratives contributes to a broader collective understanding of physical experience.

Movement-based performance is still an emerging field with few formal pathways to follow. Musicians navigating this highly personalized field require both well-developed technical and physical skills to design or adapt existing gestural systems for their own use. It is an area that encourages self-directed performance and design methods. Although musicians can draw on pre-existing software, toolkits and movement analysis methods, no external entity can act as a guide to access internal feelings and insights. By immersing themselves in somatic activities that emphasize awareness, performers can gain a greater understanding of how to experience and create music through their bodies.

14

Expanding agency

This chapter examines the impact of long-term gestural performance on performer agency. It looks at three areas: vocal, movement and general self-agency. Marc Leman (2016) defines agency as human movement executed with a goal or intention, which invokes a feeling of being in control. When movement patterns align with music, such as when dancing to music or playing air guitar, a sense of pleasure and reward can result. These positive feelings can encourage deep immersion and empowerment through musical activities. As we have seen in previous chapters, bodies are infinitely adaptable, and gestural systems are also continuously changeable. Each performance with them signals a process of transformation. Learning to move in new ways while generating music can signal even more profound empowerment, shifting musicians' understandings of themselves and expanding their expressive potential beyond dominant performance conventions in live electronic music.

How performers move on stage, interact with audiences and connect with their bodies during performance differ radically between traditional and gestural performance. Performers can explore and undermine the cultural connotations associated with the voice, signified by the markers of gender, race, ethnicity and history, reimagining their sonic imprint in relation to movement and sound with the aid of gestural systems. Not only the sound and overall timbre of the voice but also the usual modes of vocal delivery and execution can become unrecognizable in such contexts. Composer and performer Donna Hewitt (2006) treats the voice as an abstract sound for processing as she performs with modified microphone stand, the eMic, reconfiguring it to transcend the gender and cultural conditioning reflected in the female vocal sound, noting, 'Electroacoustic technologies allow us to overcome certain biological, physical and emotional limitations of natural embodied voices' (Hewitt 2006, 13). Hewitt's work explores the ability to capture and reproduce the human voice, allowing it to be removed from the body and its associated biological limitations, such as breath, pitch range, timbral quality and volume/amplitude. In this way, gestural systems can extend or expand the existing vocal instrument (Hewitt 2006, 40). This approach gives the performer a new sense of agency as they recraft their stage identity or persona (Spry 2016) beyond physical limitations and cultural expectations, in which 'electronic music promises to liberate us from biology; and the transgressive space opened up by noise and the extreme allows the abhuman to emerge' (Morgan 2014, 54).

Musicians willing to explore these new forms of performance require a fertile common ground to discuss and be supported in this work. This emerging community exists across disciplines and art forms. It is reliant on the capture and dissemination of first-person accounts that reveal how physical experiences in this realm translate to insights that solidify into new performance strategies, approaches and forms of identity. In the next section I argue that a felt connection to voice, movement and breath further reinforces agency, empowerment and transformation. When musicians feel present in their bodies, no external force can interfere with that awareness. This internalized focus can act as an anchor point for generating musical experiences through the body. This transformational process can influence gender relations and representation in musical and technical performance roles, as well as undermine restrictive notions of body normativity and the limitations of behavioural conformism.

Vocal agency

The voice is often a point of connection in a recording or live performance – particularly in popular music and opera. Lyrical content, melodic contours and phrasing unite supporting instrumental elements in song-based arrangements. In pop music instrumental parts usually follow the timing of vocal phrases. The unique sonic identity or aural imprint of the performer's voice is influenced by their physical architecture, skill and the emotions underlying the performance, forming a distinctive central focus to orient listener perception. This uniqueness makes voice an ideal partner for the individual nuances of movement.

Expanding notions of the voice and reclaiming agency of the voice through technology is possible in gestural musical performance, fulfilling Warren's (2018, 31) ideal of vocal agency as 'input into how one's own technologized vocal body mediates and is mediated'. First-person accounts of this process are rare, leaving this potential largely unacknowledged. Without this discussion about the ability to 'redefine the voice as a fundamentally agential instrument' (Warren 2018,

31) the notion of a powerful, agential female voice remains elusive. Warren sees the female voice as dispossessed and appropriated in contexts where other actors assume control over it, such as sound and mastering engineers, audio producers, software developers and composers. Yet it is possible to disrupt inbuilt biases and shift existing power relations within performance and recording in a way that acknowledges and broadcasts performers' needs and desires.

Using physical movements as an input into a gestural system, vocalists can manipulate the voice as if it were a tangible material – to loop, layer and harmonize with it. The timbre of the voice can be stretched and stripped of selective frequency content. Vocalists can transform their sound beyond physical constraints and pitch limitations, blurring markers of identity like gender and physical stature. Singers interacting with gestural systems, in redefining their movement behaviours on stage, can also develop a deeper understanding of their physical abilities and sensory experience. The patterns of everyday behaviour, which become increasingly restrictive with age if individuals maintain limited movement repertoire, can impact vocal skills and expressiveness. As Sarah Whitten (2017), voice teacher in the Holden Voice Program at Harvard University observes,

> Modern life has left us with bodies that are 'casted' by sitting, wearing shoes, and moving through a limited range of motion. The good news is, our bodies are relatively plastic and have an amazing capacity to change. We just need to know how to make the change effectively and be consistent in our efforts.

When gesture meets voice, breathing regulates vocal phrases, which in turn shapes movement phrasing. Speech also relates to phrase structure as Wishart notes in his keynote address at the 2018 Sound and Music Computing Conference (SMC), which explored the rhythmic and melodic patterns of recorded spoken phrases. Wishart's extended vocal explorations exploit the voice's ability to glide and transcend fixed pitch transitions. It is the freest of body instruments. In discovering correspondences between vocal and movement phrasing through my own work, the interplay between motion and vocalizations has often transformed into a conversation fusing spontaneous utterances and improvised upper body gestures.

Both Pamela Z and Julie Wilson-Bokowiec report increased levels of somatic awareness associated with their long-term vocal and gestural practice. Being attuned to both movement and the voice can lead to greater awareness of the involuntary processes of the body such as heartbeat and breathing. Oliveros's meditative sonic practice 'integrated voice and breath' (Bell & Oliveros 2017, 73), a method that caused her to never force her voice, enabling her to sing for long periods without strain.

Warren (2018) argues that new vocal augmenting technologies can return agency to the voice in Western music by counteracting the tendency of packaging and externalizing the 'disembodied voice' based on an ocularcentric focus on the vocalist that ignores a performer's felt connection to their voice:

> Recording technologies' early and ongoing colonisation of voice has inscribed the history and present of mediatised voice with patriarchal power dynamics, retarding the evolution of vocal agency even as cyborg agency came of age in the posthuman era. (Warren 2018, 31)

Rather than the microphone being a liberating technology of amplification, magnification and expansion, technological and performance control have become divided. When the performer arrives at the microphone and is reliant on a sound engineer for effects and blending of sounds, the transmission of their voice is outsourced and externally manipulated.

Through sound technologies that act as prostheses to the vocal body, Warren argues, vocal performers can become cyborgs and reclaim agency lost through colonization since the early years of recording in the Western tradition. The voice is involved in complex phases of what Warren calls the performative acts of the extended voice, including recording, processing and steering the resulting sonic output, where 'the embodied act of self-listening accomplishes the integration of vocal body and technological prosthesis into a new cyborg whole' (Warren 2018, 31). Techniques like looping, granularization, delays, pitch shifting and manipulation of precomposed samples can transport the voice. In digital voice practice Warren (2016, 1) observes a common divide between the composer and the performer in live electronic music:

> Existing voice-technology work frequently entails a division of labour between the composer-technologist, who creates the hardware/software, devises the formal structure, and writes about the work; and the performing vocalist, who may have some creative or improvisational input. Thus, many scientific papers on digital voice lack an authorial performance perspective.

Warren addresses this gap through her practice as a composer-technologist-vocalist. She augments vocals with her customized Abacus system. Warren regards hands as ancillary to vocal performance, instead giving the gestures of the mouth and vocal tract greater prominence. For this reason she attaches the bespoke system to the microphone to draw attention to the micro-choreography of the mouth and hands. Vocalists who fulfil the hybrid roles of designer, software developer, engineer, producer, composer and project manager are able to define their own sound at all stages of the performance and production process. These multiple identities are evident in the stage works of Pamela Z, Julie Wilson-Bokowiec, Laetitia Sonami and Lauren Sarah Hayes, who reimagine natural vocalizations through movement and technical innovation.

In-depth first-person accounts are pivotal to recognizing and cementing performer agency. Yet as Warren (2018, 32) contends, when coming up against Western associations of the voice, 'agential commentary on one's own voice is often dismissed as mere noise or ugliness'. Singer, actor and educator Shannon Holmes (2016), who conducts autoethnographic research into somatic vocal methods within rehearsal, reveals that self-examination and reflection on embodied knowledge can liberate vocal expression by reducing the focus on the mechanical aspects of vocal performance, leading to the type of integrated experience that underpins musical agency (Leman 2016). Holmes (2016, 191) identifies that the common goal of classical voice training is to produce a 'well-formed and well-co-ordinated instrument'. She finds that many of her students struggle to realize emotion in text when transitioning from spoken to sung contexts. Key aspects of reconnecting with this elusive emotion include spontaneity and vulnerability to be able to access the unknown and unfamiliar.

The crucial balance rests in balancing technique and control with these more intuitive elements. Harmony between technical, somatic and emotional states can be accessed by observing the self in action (Holmes 2016, 193) to uncover inner tensions and barriers to uninhibited vocal expression. Overly focusing on the technical aspects of performance can result in constrained, almost robotic, vocal delivery. An excess of control over the body can make it seem lifeless and unnatural, leading Holmes to conclude that 'the challenge shared by singers and actors alike is to avoid over-analysing every sound emerging from their bodies' (Holmes 2016, 194). During a voice and performance workshop with Linda Wise, Holmes (2016) was encouraged to focus on how pitches, sounds and words felt in her body rather than seeking meaning, allowing her to tap into the feelings informing technique rather than trying to control the voice through predetermined strategies.

When vocalists focus on the feelings or emotions moving through the body reflectively, freer vocal expression can occur. Rather than think how they *should* sound or *should* look on stage, the body becomes an archive of emotions and internal images available to access during performance:

> The integration of a practiced vulnerability and the use of the body as a co-performative agent to connect with one's autobiographical story may allow a space in which the body, as instrument, can experience the freedom of embodied vocal expression. (Holmes 2016, 201)

In current vocal discourse, particularly in the classical sphere, vocalists aim for perfection and control. Yet embracing the blurriness and messiness of the human body, stripped back to its raw, primal desire to produce sound that releases pent-up feelings or connects with others can expose a far deeper form of expression. By recognizing all of the activities supporting vocalization, not just the vibrating vocal cords, this misplaced endeavour for perfection is challenged (Eidsheim 2015).

The virtuosic abilities of well-trained vocalists are lauded throughout concert halls, talent shows and reality television. Vocalists have assumed and almost absorbed popular audio effects like pitch-correcting software, auto-tune, while rappers emulate stuttery, glitchy gates to blend with heavy, minimalistic rhythms. Yet gestural control enables vocalists to move one step further into the realm of interweaving performance actions with bodily sound. These very motions alter the artist's voiceprint through digital effects processing while also retraining the body to morph and mutate in novel ways during performance.

Movement and agency

Choreographer Heather Harrington (2020) links female physicality with agency and empowerment. She uses her practice to access her inner emotional world and encourages her dance students to transcend traditional female identities and assume their own process of self-discovery using Martha Graham technique. Similarly, my movement-based art and music are an extension of my deepest self-expression. The feeling-based approach I implement calls on tacit, intuitive knowledge, which is more forthcoming through immersion in free dance and improvised movement than sedentary activities like deep contemplation and writing. After a free movement or yoga session, I often receive new, unexpected insights about a current artistic or writing project. This experience is backed by studies linking physical activity with heightened creativity (Rominger et al. 2020). In favouring this inner-directed style of practice, I resist the urge to become an object of entertainment, a role that female performers, particularly vocalists, have been traditionally cast in.

Regular and varied movement is critical to maintaining health. A range of neuroscientific studies reveal the effects of physical exercise on changes to the functional and structural characteristics of the brain, enhancing neuroplasticity, which influences learning and skill acquisition (Budde et al. 2016). Recent research also demonstrates that habitual movement can contribute to a decline in mood-related illnesses such as depression (Zhao et al. 2020). Yet the popularity of high-intensity repetitive exercise is commonly viewed as a strenuous pastime involving punishment and self-deprivation for long-term gain. Faced with this disciplined form of movement, it seems easier to sink back into the safety and comfort of a sedentary lifestyle, which offers more immediate reward. The disincentive of becoming injured or exhausted during exercise often leads to a reversion to minimalistic movements associated with technology usage such as keyboard typing, thumb scrolling and screen gazing.

Sedentary daily work and recreational activities can lead to poor posture and accumulated muscular tension or weakness, eventually resulting in chronic pain conditions. For dance teacher and movement therapist Mary Whitehouse (1995, 245), these negative effects are a byproduct of being distracted from the felt sensations of the body. When movement becomes purely functional and habitual, the kinaesthetic sense can become underdeveloped, making individuals prone to habits of restriction and poor posture (Whitehouse 2007).

Whitehouse sought to elevate movement from habitual restrictions in order to cultivate an internal state of alignment rather than an ideal, external image of what a dancer should look like or what techniques they should perfect. Somatic practices such as Feldenkrais demonstrate how awareness of movement can improve economy of motion and postural alignment, so that the body is working holistically when engaging in everyday activities and musical performance. Practitioners observe areas of resistance in the musculature and set out to correct these imbalances that may restrict movement and eventually cause pain. This makes the Feldenkrais method a popular educational and therapeutic resource for musicians, who often need to assess habitual movement patterns to enhance performance, particularly under stressful conditions or during lengthy touring periods.

When analysing the agency of the performer in instrumental interaction, the notion that novel tools have the power to shape a performers' experience leading them into new, unexpected domains is common in technology discourse. Within this category, gestural systems, fed by the inputs of movement data, which carries broad individual information about personality, mood, health and expression, can also extend the agency of the performer. When the performer is making key interface decisions during design, they are integrating their body signature and knowledge into the design and customization of a tool. They are incorporating aspects of their experience in the technology and its implementation in performance, resulting in increasingly individualized original works. The habitual actions and gestural routines present in conventional musical instrumental practice differ from the open-ended, ever-changing movement vocabularies developed for highly customized gestural systems, which evolve with the musician. The player's body schema and the interface evolve together in gestural system design and performance, making it a unique case in instrumental interaction. It is this flexibility that creates new opportunities for increasing performer agency.

De Souza (2017) relates performer agency to body schema, which aids the perception of an instrument's affordances. He observes that disruptions to perceived agency are often accompanied by issues with body schema, pointing out that psychologists and philosophers distinguish between an action, or 'a *sense of ownership* (the experience of my body being mine), and a *sense of ownership* (the experience of causing and controlling action)' (De Souza 2017, 78). As a harmonica player, De Souza (2017, 79) reflects on his own performance experiences, discovering that his 'sense of agency here relies on motor agency, on my schematic responses to the harmonica's affordances'. Modified auditory feedback can affect a performer's sense of agency, he argues. Yet in Julie Wilson-Bokowiec's experience, the adaptation to varying forms of foldback in different venues caused her to resort to previously remembered vocal pitch and bodily positions. Her intensive and focused rehearsals prior to performance led to the refinement of physical actions to reliably reproduce sounds in unpredictable live situations. By reinforcing her body awareness of felt sensations relating to pitch and timbre during practice, she refined her body schema to perform effectively with the Bodycoder system even without consistent auditory feedback.

Yet novel gestural systems are not only suited to expert performers but also cater to different abilities and body types. There is no need to possess the full range of human motion to be expressive with these instruments. Significant work by Stuart Favilla, Lauren Sarah Hayes and Garth Paine in disability and dementia care demonstrates the relevance of movement-based systems to bodies that do not conform to societal norms. For individuals of limited movement range, eye gaze and micro-movements can also be channelled through gestural systems, as shown in the improvisational system for physically disabled adults designed by Allan Lem and Garth Paine (2011). In related research, Hayes assembles technology that caters to varying bodies and abilities including individuals with cochlear implants and learning difficulties. Visual, auditory and kinaesthetic feedback can be adapted to suit the abilities and preferences of the user. This increases the potential for inclusiveness and accessibility of such devices. Research within the accessible technologies sphere ensures that emerging interactive musical systems are not only created for healthy male bodies of average dimensions but also tailored to individuals with varying physical dimensions and abilities (Frid 2019).

Empowerment through movement

Acquiring advanced movement skills through regular and prolonged performance practice can have empowering impacts on dancers and other movement experts. Within the emerging neuroscience and dance field, neuroscientist and dance performer Hanna Poikonen studies the influence of movement experiences on the brain. Poikonen, Toiviainen and Tervaniemi (2018) have observed mood-related effects such as euphoria in trained dancers, as well as a greater capacity for synchronicity compared to musicians and non-expert dancers. Poikonen also found that movement can lead to a flow state by stimulating deeper brain areas. Musicians honing their movement skills while learning gestural systems experience similar effects, accessing the body as a source of knowledge. By focusing on how a movement feels rather than how it appears, kinaesthetic awareness among performers grows, informing future interactions.

Kinesics, the study of physical communication through gestures and facial expression, was invented by anthropologist, Richard Birdwhistell in 1952. It popularized body language research, exploring links between movement and personal power. Much has been written about power poses in business settings and the recruiting industry – maintaining steady eye contact, an upright posture and refraining from fidgeting are seen as essential elements of achieving success in a face-to-face job interview. For the choreography of a collaborative piece co-composed with Donna Hewitt in 2018, *#Me Too* we improvised around themes of power, influence and empowerment through physical gesture.

We experimented with expansive and contracted stances to express power relationships fluctuating between dominance and submissiveness. We echoed and interpreted each other's movements, processing our voices through each of our customized gestural systems, amplified by video projections and the matching choreography of two dancers.

My solo movement-based works also explore the influence of normative cultural restrictions on the female body. Drawing on Spry's (2016, 73) notion of performative autoethnography, which relates the personal to the political, I engage with previously held community and family values and experiences throughout performance. Sounds are mined from spontaneous sound-movement improvisations. The process has the added benefit of temporarily reversing the unwelcome side effects of prolonged experience as a sedentary studio and office worker, leading to diminished movement range and accumulated muscular tension. The effect of these ingrained behaviours on the overall expressiveness of my movement style and energetic output is a less restricted and repetitive movement vocabulary. Yet there is a positive benefit to certain repeated actions according to De Souza (2017, 18), who argues that 'habit also enables human performance'. He refers to these loopable actions as both embodied and ecological, concluding, 'They integrate hand and tool, body and world' (De Souza 2017, 19). It is therefore necessary to discern between constructive and detrimental movement habits . Through conscious attention to dominant gestural routines, musicians may be able to inhabit their bodies more purposefully, transcending habits of instrumental practice and everyday life that produce tension in the body.

15

Reimagining identity

Artists designing their own instruments assume many roles. Performer, composer and programmer Marije Baalman (2017) observed this phenomenon when reflecting on her own practice, where her various identities as artist and technologist blur together. For other musicians, composition can involve a number of contributors – a composer, programmer or instrument maker can be called in to help construct an individualized performance system. A live sound engineer, lighting operator or curator could assist in the presentation of a work. Yet Baalman benefits from knowledge in several areas, using experience in instrument-building to inform composition. She believes that the two activities are intrinsically connected: 'For computer music, one can consider the actual writing of a music program as its composition, and the code as its representation (like a score)' (Baalman 2017, 231), arguing that electronic and computer music are underpinned by '*conceptualising a system*'. The composer's identity and responsibilities have expanded in areas where art and technology intersect. The composer who is willing and able to adopt the joint roles of performer, designer, sound engineer and programmer, thrives in these environments.

Body-based activities like dancing and singing continue to be performance roles where women dominate (Bosma 2013). The exception is the purely gestural, non-sound-producing art of conducting, which continues to attract greater male participation. The majority of software developers and hardware designers are male (Tassabehji et al. 2020). Men are also employed in higher numbers than women in the technical sub-areas of computer music and sound engineering (Frid 2017). This gender imbalance is being challenged by online advocacy groups calling for greater female representation in science, technology, engineering and mathematics (STEM) fields. Finding Ada, a group named after inventor Ada Lovelace, reputed to be the first computer programmer, does this by highlighting the achievements of women in STEM.

Georg Essl (2003, 19) regards live interfaces as a gendered metaphor in his New Interfaces for Musical Expression (NIME) conference paper drawing attention to the issue of under-representation of women in the field of novel digital instruments. His main argument indicates that the intersection of bodies and technology in NIME makes it a fertile area for examining gender imbalances in technical areas, postulating 'that this technology is of particular interest for the artistic exploration of gender by making the body the recipient of performance':

> This technology inverts the role of the body. Instead of being the site of control and performance, it becomes the site of reception and audience. This technology illustrates how new music technology can further contribute to gender discourse through this inversion for it addresses the relationships between body, gender and sex.

Adopting a post-structuralist approach, Essl outlines the importance of questioning persistent binaries such as woman/man, composer/performer, and nature/technology. This theme remains present throughout a range of works on gender and music technology. Essl draws on the feminist perspective of Judith Butler (1990), who recognizes the strong link between the performative and the body and gender, questioning why these dualities exist and how they are reproduced in everyday life.

Although he hesitates to form generalizations about a typical male or female sound, Essl (2003, 26) compares and contrasts performances by Waisvisz and Sonami, noting that Waisvisz's style is harsh, aggressive and mechanistic, while Sonami's is softer, more subtle and nuanced. Yet Tanaka's performance *Tibet*, featuring Tibetan singing bowls, also takes a gentler approach, contradicting the formation of sound-based gender stereotypes. Essl then shifts to exploring the differences between women and men performing in the field, asking: 'Are new music technology performances by women more likely to be gendered, gender aware, gender critical, or gender deconstructive? If so, why are they? Why, then, are performances by men rather comparatively ungendered, gender unaware, or uncritical?' (Essl 2003, 26) Interviews with female artists, Pamela Z, Laetitia Sonami and Julie Wilson-Bokowiec reveal high levels of gender awareness, reinforced by long-term performance experiences causing them to question gender roles relating to voice, dance and technology. Yet Stuart Favilla also debates the patriarchal foundations of computer music and composition, demonstrating that a critical perspective and awareness of gender divisions in technological performance is not exclusive to female practitioners. These first-person narratives by both women and men performing and designing performance systems offer a significant contribution to understanding current gender conventions and imbalances in the field.

An analysis of New Interfaces for Musical Expression (NIME) papers by Essl (2003) early in its history revealed that far fewer women participated in authorship than men. There was also an under-representation of sociologists, ethnographers and gender theorists. The structuring of conference papers as technical reports may contribute to this gap. The common template of NIME, International Computer Music Conference (ICMC) and Sound and Music Computing Conference (SMC) papers follows a formal scientific paper format in which the background research usually leads up to a technical description of the functionality and mapping of a custom-designed interface, a trend that continues in recent NIME proceedings (Hayes & Marquez-Borbon 2020). A smaller part of the ensuing discussion might mention relevant social aspects of the interface's application in performance; however, this is usually placed in a subordinate position to the technical considerations of a research project or creative work.

There are also notable differences around expectations relating to sound and gender, as outlined by Marie Thompson in a special issue of *Contemporary Music Review* in 2016 dedicated to exploring the politics of gender in regard to digital music, creative practice, higher education, computer music, electronic music and sound art with Sonami on the cover. Thompson (2016) argues that feminized musical styles such as pop are often dismissed as superficial, superfluous and brash, taking up excessive frequency space and signifying lowbrow content. Women are rewarded not only for taking up less space and moving more discreetly than men (Young 1980) but also for not raising their voices excessively or over-aggressively and making less noise overall. Female performers like Courtney Love and Britney Spears are 'outed' by sound engineers who release dry, unaffected recordings of live vocal performances out of context, shaming them for wandering outside the realms of acceptable technical vocal ability. Yet this type of public shaming rarely occurs for male artists. Women musicians have also reported that it can take longer to communicate their technical stage set-up to a male sound engineer and gain understanding and agreement (Naphtali 2017), confounding the situation further.

Thompson (2016) outlines a range of stereotypes that females are typecast in for the unacceptable noise they create, including the uncontained madwoman, seductive siren and toxic twitter feminist who dares to resist contradictory media norms and is often silenced by trollish abuse. Female gamers and software developers have been sexually harassed and intimidated on technical forums for daring to occupy the internet with their ideas. In line with Thompson's observations, I have experienced a range of dismissive and pornographic slurs in 'respected' online audio-technical communities, causing me to reduce my overall contribution in such spaces. However, awareness of gender issues in technical forums is increasing. After initial resistance, a popular music technology forum changed its name in 2021 from the derogatory label Gearslutz to Gearspace after an online petition signed by 5,000 people argued about the alienating and unprofessional gender bias implied in the name.

The preconceptions underlying sound-based stereotypes aligning madness, physicality and seduction with female identity are evident in the exclusion of women from the history of sonic experimentalism that includes compositional innovations in noise music. The songs of female band Le Tigre were critiqued and trivialized for lacking sonic impact and structure, yet their male counterparts' glitches and mistakes were interpreted as intentional and innovative, Bosma (2016) found. Female experimentalists have also been largely written out of technology-centred noise music narratives according to Bosma. This dismissiveness translates into a blanket exclusion of women composers and performers from historical accounts of noise music (Bosma 2016).

Women's higher participation in vocal and dance roles, displaying skills that are often seen as feminine and disregarded as 'soft' and 'non-technical', contributes to the partial exclusion of female performers from sound art discourse. The research field is also male dominated, constituting a type of 'techno-fetishism' (Rogers 2010, 7) in which mastery over machines is paramount. Various collectives have been formed to address persistent unequal gender representation in electronic music and sound art, including Tara Rogers' web-based activist project, Pink Noises, which was later turned into a book of the same name (Rogers 2010). In related work, feminist musicologists Sally McArthur (2014) and Frances Morgan (2017) look beyond existing histories of composers and technologists to discover how pioneers are defined and recognized.

In deciding to truly inhabit my body, I've attempted to reclaim it from social norms, incorporating lessons from this research. I've embraced embodied and performative autoethnography to reflect on musical and movement-based experiences in the vein of Tami Spry (2017, 51), who states: 'It is autoethnography that activates the foundational sociocultural personally political reflexivity of that body/self.' Yet self-definition is not sufficient. Relationships with others, as much as self-reflection, Spry asserts, contribute to definitions of identity: 'Rather than constructing identity, autoethnography is about articulating the relational effects of our differences' (2017, 51). Spry (2017, 52) reflects on Haraway's visions of hybrid futures in which she asks 'Who are we?' rather than 'Who am I?' to arrive at a type of embodiment that recognizes the cultural influences that influence individual behaviour.

Interactions with the audience can deliver new insights for musicians. Even indifference provides useful information. Kinaesthetic empathy connects performer and audience, just as novel instruments offer instructive personal data about bodily states, conditions and emotions. Gestures offer a shared cultural language that reflect prevailing messages about culturally accepted notions relating to sexuality, gender, race and power. Gestural systems offer vocalists the ability to remotely shape their own sound, thus empowering them to electronically mediate their voices and mix music with minimal involvement from technical personnel. Assuming hybrid identities that blend the technical and organic, female vocalist-composers like Pamlea Z, Julie Wilson-Bokowiec and Lauren Sarah Hayes merge traditionally masculine areas of composition and technology with conventionally feminine performance roles (Bosma (2013, 218). Building on Green's (1997) research highlighting the alignment of technology and improvisation with masculinity, Bosma demonstrates that these performer/composers, in also adopting solo instrumental or vocal performance, which is typically aligned with femininity, hybridize these established gender identities and move fluidly between them.

Future directions

Increasing exposure to gestural systems through mobile phone use and public touchscreens offers greater opportunities for developing self-awareness and the potential for identity renewal, becoming potent tools in political transformation. When the body truly becomes the instrument, absorbed into a symbiotic technological fusion with the machine, it becomes a form of transcendence:

> As a performance strategy, it blurs the notion of control by the player over the instrument, establishing a different relationship among them, one in which performer and instrument form a single technological body, articulated in sound and music. (Tanaka & Donnarumma 2019, 94)

The body is continuously reimagined in gestural performance contexts. The implications of this emerging form of the cyborg as 'a creature in the post-gender world' (Haraway 1991, 457) are still unfolding. Current technological trends in the field are moving towards greater diversity of human involvement. Emerging first-person narratives reveal the unique felt experiences of artists using novel gestural systems. Coupled with innovations in augmented and virtual reality, artists crafting and playing gestural instruments are actively exploring the dissolution of performer/audience boundaries.

Participation in gestural performance can encourage beneficial physical engagement in musical creation, offering a sense of immersion and play. These positive feelings can in turn create the motivation to create more nuanced and varied movement experiences. The kinaesthetic abilities of the performer are activated through new forms of awareness of the internal sensations of the body in motion through playful exploration of sound generation and manipulation with gestures and continuous body motion.

As previously discussed, a transparent connection between gestural instruments and the body ensures that they are more easily absorbed into the body schema, potentially promoting greater user satisfaction and broader uptake. The benefits of addressing persistent design challenges by focusing on the first-person experience and skills of performers can lead to gestural systems that sharpen the senses and movement awareness. Musicians engaging with gestural systems are tasked with learning to move through space more consciously and inhabiting their bodies more deliberately. Developing enhanced movement mastery allows musicians to deliver added personal character to performances with technology and vary whole body movements so that they can be freed from repetitive and restricted postures that hamper full expression. Playing gestural systems challenges performers to discover new movement vocabularies as a means to generating novel sounds.

Reflection and embodied autoethnographic writing continues to play a significant role in nurturing self-awareness. Writing has become a vital act for processing sensations and emotions for me during post-performance and recording. As Linda Candy (2020, 187) observes,

> Practitioners engaged in embodiment research through art have distinguished between sensory perceptions that are understood cognitively (e.g. saying 'it reminds me of when …') from those which are embodied (e.g. 'it feels as if I am inside the womb'). Where these are prompted by image sensations, the first are

associated with memories or past experience whilst the second can be deeply felt so as to be transformative.

Analysing felt sensations in written form offers access to body knowledge that might continue to be submerged or hidden otherwise. Writing places felt sensations in the light, where they can be acknowledged and examined, though there is a danger of over-dissecting or over-analysing these raw inner feelings by verbalizing and recording them as text. Candy interviewed a group of practitioners incorporating reflection in embodied interaction design and performance, including dancers Sue Hawksley and Sarah Fdili Aaloui and interaction designer George Khut, revealing diverse approaches to the body and movement-based interaction. Each artist yields insights into the role of felt experience in body-centred works. Dancer and choreographer Hawksley approaches her main medium, dance, as a path to implementing the philosophies of embodied cognition and knowledge through her embodied creative practice (Candy 2020, 210). She is particularly interested in how the interaction between people and technology influences the way they sense, feel and act. Turning inward through meditation also becomes a way for the audience to appreciate the performer's experience from a different perspective.

Khut's works reveal that attention to inner body states through focus on breath and heartbeat can enhance bodily experience. His software has been used in therapeutic environments, like helping children to turn inward and relax, rather than forming a distraction, before and during medical procedures. Fdili Alaoui considers technology as a partner and extension of the body in her research, and is also dedicated to accessing the knowledge that emerges from 'felt movement'. These tools balance somatic awareness with programs and mappings that are self-created or borrowed from friends and collaborators. Fdili Alaoui accesses a toolkit with electromyogram (EMG), proximity and Inertial Measurement Unit (IMU) sensors, choosing a particular tool for a specific context. She also uses her skills in computational creativity to implement machine learning. Yet these tools are not placed at the forefront of her design practice, but rather act as extension of the body, she contends, 'I think you need a simple connection to the body. The easiest way is to have an embodied practice' (Fdili Alaoui, cited in Candy 2020, 216).

First-person narratives of women and gender diverse artists in live electronic music and movement-based performance assist in redressing their under-representation in electroacoustic and noise music research. The substantial contributions of female performers and programmers including Lauren Sarah Hayes, Laetitia Sonami, Marije Baalman and Rebecca Fiebrink illustrate creative links between music creation and coding, highlighting the need for more women to be recognized in the design and software development aspects of the gestural performance field as an example for students and musicians entering it. As Magnusson (2019, 229) points out, 'A relatively small band of actors in the Bay Area, Berlin, and Stockholm are largely responsible for how our musical culture develops, with some further research taking place in Paris, London, and other cities.' The concentration of audio software development in a few limited regions and institutions in turn affects the diversity of emerging innovations.

The redirection and redistribution of artificial binaries of mind/body, human/machine, manufactured/natural, masculine/feminine are a prerequisite for the dissolution of power and representational imbalances in the highly technical gestural performance field, which Bosma (2013) links to the under-reporting of women's achievements in electroacoustic music. She addresses a further trend among female composer/performers towards interdisciplinarity, which can also obscure women's contributions and broader acknowledgement in the area. The interdisciplinary nature of women's work in the electroacoustic music field places them at danger of being ignored or under-represented in historical narratives, Bosma warns. This diminished visibility can impede funding opportunities, as many grant categories concentrate on one art form, making it challenging for women to position themselves for optimum success:

> Traditionally, composing, improvising, and technology are male-gendered domains, while singing and performing specific solo instruments are gendered as feminine. Women composer-performers disrupt this stereotypical dichotomy. The traditional notion of authorship may also collapse through women's preference for interdisciplinary work and collaboration: then authorship may be shared among several artists and the traditional roles of composer and performer unsettled. (Bosma 2013, 233)

Women often flit between the gaps, between art forms – a classic characteristic of movement-based systems in which data is fluid and interchangeable. Gravitating towards group and interdisciplinary projects can dilute women's contributions, spreading them across a range of focus areas, particularly in collaborative audiovisual performances that straddle theatrical and dance contexts.

As individuals hybridize with technology they assume neutral gender identities, according to the cyborg vision mapped out by Haraway (1991). Gradually, the dualisms that Haraway and other post-structuralists identified are disintegrating. This transition feeds into the emerging discourse treating technology as a pathway for feminist resistance. Coding rights directress of the site www.transfeministech.org Joana Varon developed a transfeminist oracle card deck in association with media scholars Sasha Constanza-Chock and Clara Juliano to playfully envision new technological realities unconstrained by heterenomic rules and standards. Responding to the binary and normative nature of technology, the alternative values the researchers explore include agency, autonomy, empathy, embodiment, intuition, pleasure and decolonization.

Reacting to privacy breaches and permissiveness of violent and racist forms of 'free speech' on social media, Varon asserts that values, context and the individuals who create technology all matter. The contributions of culturally diverse designers and performers to creating hybrid and bespoke gestural systems contributes to increasingly inclusive systems suitable for a range of body types and preferences. As part of this innovation, Varon sees imagination as an instrument for revolution. Designers need to envision an alternative future to replace outdated and dismantled social conventions. The transfeminist oracle card deck invites contemplation of ethical and non-competitive forms of technological development that promote self-agency and reimagine dominant technical imperatives of surveillance and external control.

Alternative values are also explored in gestural system performance. As Kazlauskaite (2020) argues about the sound art world, gendered language and body inhibition can prevent women from fully realizing their creative potential as well as being recognized as artists. Many female practitioners have discovered a new sense of empowerment through whole body expression in gestural performance, like the increased movement fluidity Pamela Z experienced through her ongoing engagement with the BodySynth. Compared to the jittery and robotic motions of her initial performances, a new intricate and nuanced movement vocabulary developed as her work with gestural systems matured.

The process of creating dedicated repertoire for voice and gestural systems and also translating this essentially ephemeral practice to the recorded medium offers an alternative method of developing gestural performance practice. However, several artists interviewed have expressed concern that the innate character of this ethereal and intangible art form may be lost when captured onto a recorded medium, prompting the question of whether potential audiences need to attend gestural performances in person to sample their liveness. Sonami considers all sides of the debate in relation to her practice:

> I don't have many recordings because for me, much of the gesture and the performance is about just being with people. … For instance, when these people ask to film the performance, I'm like no way, I don't do that stuff, I'm just a performer, I'm just about presence. Then I talked to a friend who said why not do something you could never do otherwise? Why not think of something you could never do when you were touring? I think that if you record, it's almost like a gig because of the instrument that you built. In performance things are happening so differently than when controlling with Pro Tools. Things are happening and you're hearing, so it's almost like you just have to do that and completely be in the performance mode and then after that it's a recording devoid of any presence. But it happened because of presence so it's already informed by presence.

Recordings allow musicians to capture snapshots of particular moments of their instrument's development. They enable performers to indulge in more complex and considered set-ups than afforded by touring, offering opportunities for further layering and gestation. It is the challenge of the performer to summon a similar level of energy and intensity within a recording as in a performance before a live audience that generates its own group energy, dynamism and unpredictability.

Warren (2018, 32) recommends that 'future work should delve more deeply into the spectrum between liveness and recording in extended voice practice'. This suggestion can also apply to gestural performance practice, where recording can contribute to solidifying and disseminating ideas beyond the largely specialist audiences of experimental live electronic music residing in concert halls and laboratories. Recordings of original works spread the sounds of embodiment to a wider audience, allowing artists to capture and analyse their progress in developing ideas for emerging and fleeting gestural systems and instruments. Tanaka remembered hearing a recording of a Waisvisz performance with *The Hands* on a musical compilation, reflecting on the physicality embedded in the performer's sound. Even without video, it was possible to sense the presence of the body in the music:

> Feeling the real viscerality of the music showed me the recording that was so distinctive and different from the other pieces on that compilation that were

more studio works. Then I actually realised that the physicality of gesture can survive the recorded media.

The experience of hearing individual bodily inflections imprinted onto sound ultimately inspired Tanaka to enter into gestural music as a student, even without witnessing the gestures behind the sounds. When the body of a performer is inscribed in a sound, it is imbued with all their unique emotions, habits and creativity.

Yet there is a dilemma associated with recording gestural systems that may cause audiences to question the authenticity of the live performance, according to Wilson-Bokowiec:

> I think that the thing about the recorded medium is that it plays to some of those unfounded presumptions about interactive technology, in that if you say that everything is live, everything is coming from the voice, I'm sampling the voice live at this moment in time, people don't understand. They presume there's some kind of background score going on, that the computer's playing something. They don't understand that everything is coming from me – the voice, the gestures. So when you then start offering recordings of that, it then plays into this notion, well it's a recording anyway. It's an interesting catch-22.

This observation reflects an ongoing debate about authenticity in live electronic music where connections between performer actions and sound are less obvious than in acoustic instrumental music. Moving between live and electronic media can colour audience expectations. However, existing recorded repertoire with gestural systems, which extends to pop music by Imogen Heap and the iconic use of the theremin by The Beach Boys, in the BBC radiophonic workshop and other cult contexts, can benefit from the addition of further recordings of works featuring gestural instruments designed by the performers who play them. A broader body of original recorded material in the field can increase awareness of movement-centred performance practice beyond research and academic contexts.

Linking bodily motion and sound can imprint the body on a musical work. No body moves quite like another. Bodies offer an imprint of individual personalities, acquired skills and past experiences with sound and movement. Sounds produced with gestural systems are infused with the embodied personality and experience of an individual. Yet this emerging type of performance entails risk and a degree of bravery. Gestural systems promote customized approaches in live electronic music – away from the constraints of packaged commercial audio software and equipment. They prompt musicians to act in exploratory and unexpected ways in public settings without the safety net of a cleverly engineered instrumental object to orient their actions. The process involves contradicting preconceptions and cultural conditioning, which from an early age instruct individuals how to occupy their bodies and safely propel them through space in accordance with a predesignated social identity. Through gestural performance musicians have the potential to share unique movement expression. Playing gestural instruments involves directing energy and inner feelings. Performers can make the intangible tangible through their imagination and physical contributions to music-making and sound art. By placing the body and movement awareness at the centre of performance, gestural systems offer musicians a heightened sense of immediacy and immersion, allowing the body to become a conduit for sounds otherwise inaccessible.

References

Acitores, A. P. (2011), 'Towards a Theory of Proprioception as a Bodily Basis for Consciousness in Music', in D. Clarke & E. Clarke (eds), *Music and Consciousness: Philosophical, Psychological, and Cultural Perspectives*, 215–30, Oxford: Oxford University Press.

Andersen, K., & Gibson, D. (2017), 'The Instrument as the Source of New in New Music', *Design Issues*, 33(3), 37–55.

Andreallo, F. (2019), 'The Selfie Generation: A Transformation of Visual Social Relationships', *Vista*, 4, 153–71.

Antle, A. N., Corness, G., & Droumeva, M. (2009), 'Human-Computer-Intuition? Exploring the Cognitive Basis for Intuition in Embodied Interaction', *International Journal of Arts and Technology*, 2(3), 235–54.

Avila, J. M., Tsaknaki, V., Karpashevich, P., Windlin, C., Valenti, N., Höök, K., McPherson, A., & Benford, S. (2020), 'Soma Design for NIME', in *Proceedings of the International Conference on New Interfaces for Musical Expression*, 489–94, Birmingham, UK: Birmingham City University.

Baalman, M. A. (2017), 'Interplay between Composition, Instrument Design and Performance', in H. Egermann, S. I. Hardjowirogo, S. Weinzierl, T. Bovermann & A. de Campo (eds), *Musical Instruments in the 21st Century: Identities, Configurations, Practices*, 225–41, Singapore: Springer.

Behnke, E. A. (1995), 'Matching', in D. H. Johnson (ed.), *Bone, Breath and Gesture: Practices of Embodiment*, 317–37, Berkeley, CA: North Atlantic Books.

Bell, G., & Oliveros, P. (2017), 'Tracing Voice through the Career of a Musical Pioneer: A Conversation with Pauline Oliveros', *Journal of Interdisciplinary Voice Studies*, 2(1), 67–78.

Bencina, R. (2005), 'Metasurface: Applying Natural Neighbor Interpolation to Two-to-Many Mapping', in *Proceedings of the International Conference on New Interfaces for Musical Expression (NIME'05)*, 10–14, Vancouver, Canada: University of British Columbia.

Bencina, R., Wilde, D., & Langley, S. (2008), 'Gesture ≈ Sound Experiments: Process and Mappings', in *Proceedings of the International Conference on New Interfaces for Musical Expression*, 197–202, Genova, Italy: University of Genova.

Bevilacqua, F., Müller, R., & Schnell, N. (2005), 'MnM: A Max/MSP Mapping Toolbox', in *Proceedings of the International Conference on New Interfaces for Musical Expression*, 85–8, Vancouver, Canada: University of British Columbia.

Bevilacqua, F., Schnell, N., & Fdili Alaoui, S. (2011), 'Gesture Capture: Paradigms in Interactive Music/Dance Systems', in G. Klein & S. Noeth (eds), *Emerging Bodies: The Performance of Worldmaking in Dance and Choreography*, vol. 21, 183–93, Beilefeld: Verlag.

Birdwhistell, R. L. (1952), *Introduction to Kinesics: An Annotation System for Analysis of Body Motion and Gesture*, Washington, DC: Department of State, Foreign Service Institute.

Blom, L. A., & Chaplin, L. T. (1988), *The Moment of Movement: Dance Improvisation*, Pittsburgh, PA: University of Pittsburgh Press.

Bokowiec, M. A. (2011), '*VOCT (Ritual)*: An Interactive Vocal Work for Bodycoder System and 8 Channel Spatialization', in *Proceedings of the International Conference on New Interfaces for Musical Expression*, 40–3, Norway: University of Oslo.

Bongers, A. J. (1998), 'Tactual Display of Sound Properties in Electronic Musical Instruments', *Displays*, 18(3), 129–33.

Bongers, B. (2000), 'Physical Interfaces in the Electronic Arts', in M. M. Wanderley & M. Battier (eds), *Trends in Gestural Control of Music*, 41–70, Paris, IRCAM.

Bosma, H. (2013), *The Electronic Cry: Voice and Gender in Electroacoustic Music*, Amsterdam: University of Amsterdam.

Bosma, H. (2016), 'Gender and Technological Failures in Glitch Music', *Contemporary Music Review*, 35(1), 102–14.

Budde, H., Wegner, M., Soya, H., Voelcker-Rehage, C., & McMorris, T. (2016), 'Neuroscience of Exercise: Neuroplasticity and Its Behavioral Consequences', *Neural Plasticity*, 2016, 3643879. doi:10.1155/2016/3643879.

Butler, J. (1990), *Gender Trouble: Feminism and the Subversion of Identity*, New York: Routledge.

Buxton, B. (2007), *Sketching User Experiences: Getting the Design Right and the Right Design*, Burlington, MA: Morgan Kauffman.

Cadoz, C. (1988), 'Instrumental Gesture and Musical Composition', in *Proceedings of the International Computer Music Conference (ICMC)*, 1–12, Cologne: ICMA.

Cadoz, C. & Wanderley, M. M. (2000), 'Gesture-Music', in M. M. Wanderley & M. Battier (eds), *Trends in Gestural Control of Music*, 71–94, Paris: IRCAM.

Cranny-Francis A. (2013), *Technology and Touch*, London: Palgrave Macmillan.

Candy, L. (2020), *The Creative Reflective Practitioner: Research through Making and Practice*, Adbington, Oxon: Routledge.

Cannon, J., & Favilla, S. (2012), 'The Investment of Play: Expression and Affordances in Digital Musical Instrument Design', in *Proceedings of the International Computer Music Conference (ICMC)*, 459–66, Ljubljana: ICMA.

Caramiaux, B. 2014, 'Motion Modeling for Expressive Interaction: A Design Proposal Using Bayesian Adaptive Systems', in *Proceedings of the International Workshop on Movement and Computing (MOCO'14)*, 76–81, Paris: IRCAM.

Cascone, K. (2002), 'Laptop Music-Counterfeiting Aura in the Age of Infinite Reproduction', *Parachute*, 107, 52–9.

Castellano, G., Bresin, R., Camurri, A., & Volpe, G. (2007), 'Expressive Control of Music and Visual Media by Full-body Movement', in *Proceedings of the 7th International Conference on New Interfaces for Musical Expression (NIME' 07)*, 390–1, New York: ACM.

Coniglio, M. (n.d.), 'Isadora'. Computer Software. *Troikatronix*, https://Troikatronix.com/.

Coniglio, M. (2005), 'The Importance of Being Interactive', in G. Carver & C. Beardon (eds), *New Visions in Performance*, 5–12, London: Routledge.

Cook, O. (2004), *Singing with Your Own Voice*, London: Nick Hern Books Limited.

Cook, P. (2002), *Real Sound Synthesis for Interactive Applications*, Natick, MA; A K Peters.

Csikszentmihalyi, M. (1996), *Creativity: Flow and the Psychology of Discovery and Invention*, New York: Harper Collins.

Cusick, S. G. (1994), 'Feminist Theory, Music Theory, and the Mind/Body Problem', *Perspectives of New Music*, 32(1), 8–27.

Dahl, S., & Friberg, A. (2007), 'Visual Perception of Expressiveness in Musicians' Body Movement', *Music Perception*, 24(5), 433–54.

Davidson, J. W. (2001), 'The Role of the Body in the Production and Perception of Solo Vocal Performance: A Case Study of Annie Lennox', *Musicae Scientiae*, 5(2), 235–56.

Davidson, J. W. (2012), 'Bodily Movement and Facial Actions in Expressive Musical Performance by Solo and Duo Instrumentalists: Two Distinctive Case Studies', *Psychology of Music*, 40(5), 595–633.

De Souza, J. (2017), *Music at Hand: Instruments, Bodies, and Cognition*, Oxford: Oxford University Press.

Dewey, J. ([1934] 2005), *Art as Experience*, New York: Pedigree Books.

Dobrian, C., & Koppelman, D. (2006), 'The "E" in NIME: Musical Expression with New Computer Interfaces', in *Proceedings of the International Conference on New Interfaces for Musical Expression*, 277–82, Paris: IRCAM.

Doğantan-Dack, M. (2011), 'In the Beginning Was Gesture: Piano Touch and an Introduction to a Phenomenology of the Performing Body', in A. Gritten & E. King (eds), *New Perspectives on Music and Gesture*, 243–65, Farnham: Ashgate.

Dourish, P. (2004), *Where the Action Is: The Foundations of Embodied Interaction*, Cambridge, MA: MIT Press.

Eidsheim, N. S. (2015), *Sensing Sound: Singing and Listening as Vibrational Practice*, Durham, NC: Duke University Press.

Emmerson, S. (2007), *Living Electronic Music*, Aldershot: Ashgate.

Erkut, C., & Dahl, S. (2017), 'Embodied Interaction through Movement in a Course Work', in *Proceedings of the International Conference on Movement Computing*, 1–8, London: University of Surrey.

Essl, G. (2003), 'On Gender in New Music Interface Technology', *Organised Sound*, 8(1), 19–30.

Favilla, S. (1996), 'Non-linear Controller-Mapping for Gestural Control of Gamaka', in *Proceedings of the International Computer Music Conference*, 89–92. International Computer Music Association, Hong Kong: Michigan Publishing.

Favilla, S., Cannon, J., & Greenwood, G. (2005), 'Evolution and Embodiment: Playable Instruments for Free Music', in *Proceedings of the International Computer Music Conference (ICMC)*, Barcelona, Spain: Michigan Publishing.

Favilla, S., & Cannon, J. (2006), 'Fetish: Bent Leather's Palpable, Visceral Instruments and Grainger', *Contemporary Music Review*, 25(1–2), 107–17.

Feisst, S., & Paine, G. (2020), 'Sonic Intimacies: The Sensory Status of Intimate Encounters in 3-D Sound Art', in S. Krogh Groth & H. Schulze (eds), *The Bloomsbury Handbook of Sound Art*, 213–20, London: Bloomsbury Academic.

Feldenkrais, M. (1972), *Awareness through Movement: Health Excercises for Personal Growth*, New York: Harper & Row.

Feldman, M. (2015), *The Castrato: Reflections on Natures and Kinds*, Berkeley: University of California Press.

Ferguson, J. R. (2013), 'Imagined Agency: Technology, Unpredictability, and Ambiguity', *Contemporary Music Review*, 32(2–3), 135–49.

Ferguson, J. R. (2016), 'Michel Waisvisz: No Backup/Hyper Instruments', in S. Emmerson, & L. Landy, (eds), *Expanding the Horizon of Electroacoustic Music Analysis*, 247–65, Cambridge: Cambridge University Press.

Fdili Alaoui, S., Schiphorst, T., Cuykendall, S., Carlson, K., Studd, K., & Bradley, K. (2015), 'Strategies for Embodied Design: The Value and Challenges of Observing Movement', in *Proceedings of the 2015 ACM SIGCHI Conference on Creativity and Cognition*, 121–30, Glasgow, United Kingdom: ACM.

Fdili Alaoui, S., Henry, C., & Jacquemin, C. (2014), 'Physical Modelling for Interactive Installations and the Performing Arts', *International Journal of Performance Arts and Digital Media*, 10(2), 159–78.

Fiebrink, R., & Sonami, L. (2020), 'Reflections on Eight Years of Instrument Creation with Machine Learning', in *Proceedings of the International Conference on New Interfaces for Musical Expression*, 232–42, Birmingham, UK: Birmingham City University.

Fogtmann, M. H., Fritsch, J., & Kortbek, K. J. (2008), 'Kinesthetic Interaction: Revealing the Bodily Potential in Interaction Design', in *Proceedings of the 20th Australasian Conference on Computer-Human Interaction: Designing for Habitus and Habitat*, 89–96, New York: ACM.

Foster, S. L. (1986), *Reading Dancing: Bodies and Subjects in Contemporary American Dance*, Berkeley: University of California Press.

Fowler, G. A., & Kelly, H. (2020), 'Amazon's New Health Band Is the Most Invasive Tech We've Ever Tested', December 10, available online: https://www.washingtonpost.com/technology/2020/12/10/amazon-halo-band-review/ (accessed February 2021).

Frid, E. (2017), 'Sonification of Women in Sound and Music Computing: The Sound of Female Authorship in ICMC, SMC and NIME Proceedings', in *Proceedings of the International Computer Music Conference (ICMC)*, 233–8, Shanghai Conservatory of Music, Michigan Publishing.

Frid, E. (2019), 'Accessible Digital Musical Instruments – a Review of Musical Interfaces in Inclusive Music Practice', *Multimodal Technologies and Interaction*, 3(3), 57.

Frid, E., Elblaus, L., & Bresin, R. (2019), 'Interactive Sonification of a Fluid Dance Movement: An Exploratory Study', *Journal on Multimodal User Interfaces*, 13(3), 181–9.

Gallagher, S. (2005), *How the Body Shapes the Mind*, New York: Oxford University Press.

Gelineck, S. (2012), 'Exploratory and Creative Properties of Physical-Modeling-Based Musical Instruments', PhD diss., Aalborg University, Copenhagen.

Gendlin, E. T. (1996), *Focusing-oriented Psychotherapy: A Manual of the Experiential Method*, New York: Guilford Press.

Gibson, J. J. (1986), *The Ecological Approach to Visual Perception*, New York: Taylor & Francis.

Gilman, M. (2019), 'The Science of Voice and the Body', in Y. Kim & S. L. Gilman (eds), *The Oxford Handbook of Music and the Body*, 62–78, Oxford: Oxford University Press.

Gladwell, M. (2008), *Outliers*, Boston, MA: Little, Brown.

Glennie, E., Gilman, S. L., & Kim, Y. (2018), 'Is There Disabled Music?: Music and the Body from Dame Evelyn Glennie's Perspective', in Y. Kim & S. L. Gilman (eds), *The Oxford Handbook of Music and the Body*, 318–30, Oxford: Oxford University Press.

Godøy, R. I. (2011), 'Sound-Action Chunks in Music', in J. Solis & K. Ng (eds), *Musical Robots and Interactive Multimodal Systems*, 13–26, Berlin: Springer.

Grainger, P. ([1936] 2002), 'Free Music No. 1 (For Four Theremins)', *An Anthology of Noise & Electronic Music*, Vol. 2 (Disc 2), Bruxelles, Belgium: Sub Rosa.

Green, L. (1997), *Music, Gender, Education*, New York: Cambridge University Press.

Green, O. (2014), 'NIME, Musicality and Practice-led Methods', in *Proceedings of the International Conference on New Interfaces for Musical Expression*, 1–6, Goldsmiths, University of London.

Halmrast, T., Guettler, K., Bader, R., & Godøy, R. I. (2010), 'Gesture and Timbre', in R. I. Godøy & M. Leman (eds), *Musical Gestures: Sound, Movement, and Meaning*, 183–211, New York: Routledge.

Han, J. G. (2019), 'The Somaesthetics of Musicians: Rethinking the Body in Musical Practice', *Journal of Somaesthetics*, 5(2), 41–51

Hanna, T. (1988), *Somatics: Reawakening the Mind's Control of Movement, Flexibility and Health*, Reading, MA: Addison-Wesley.

Hansen, L. A. (2011), 'Full-body Movement as Material for Interaction Design', *Digital Creativity*, 22(4), 247–62.

Haraway, D. ([1985] 1991), 'A Cyborg Manifesto: Science, Technology and Socialist-Feminism in the Late Twentieth Century', in *Simians, Cyborgs and Women: The Reinvention of Nature*, 149–82, New York: Routledge.

Harrington, H. (2020), 'Female Self-Empowerment through Dance', *Journal of Dance Education*, 20(1), 35–43.

Harrison, S., Tatar, D., & Sengers, P. (2007), 'The Three Paradigms of HCI', in *Proceedings of the Alt. Chi. Session at the SIGCHI Conference on Human Factors in Computing Systems*, 1–18, San Jose, CA.

Hayes, L. (2014), 'Audio-Haptic Relationships as Compositional and Performance Strategies', Phd diss., University of Edinburgh, Scotland.

Hayes, L. (2019a), 'Beyond Skill Acquisition: Improvisation, Interdisciplinarity, and Enactive Music Cognition', *Contemporary Music Review*, 38(5), 446–62.

Hayes, L. (2019b), 'PARIESA: Practice and Research in Enactive Sonic Art', in *Mobile Brain-Body Imaging and the Neuroscience of Art, Innovation and Creativity*, 53–60, Cham: Springer.

Hayes, L., & Marquez-Borbon, A. (2020), 'Nuanced and Interrelated Mediations and Exigencies (NIME): Addressing the Prevailing Political and Epistemological Crises', 428–33, Birmingham, UK: Birmingham City University.

Heiddeger, M. ([1927] 2010), *Being and Time*, Albany, NY: Suny Press.

Hewitt, D. G. (2006), 'Compositions for Voice and Technology', PhD diss., University of Western Sydney, Australia.

Höök, K., Caramiaux, B., Erkut, C., Forlizzi, J., Hajinejad, N., Haller, M., & Tobiasson, H. et al. (2018), 'Embracing First-Person Perspectives in Soma-based Design', *Informatics*, 5(8), 1–26.

Höök, K., Jonsson, M. P., Ståhl, A., & Mercurio, J. (2016), 'Somaesthetic Appreciation Design', in *Proceedings of the 2016 CHI Conference on Human Factors in Computing Systems*, 3131–42, New York: ACM.

Holmes, S. (2016), 'Autoethnography and Voicework: Autobiographical Narrative and Self-reflection as a Means towards Free Vocal Expression', *Voice and Speech Review*, 10(2–3), 190–202.

Howard, D. M., & Rimmell, S. (2004), 'Real-Time Gesture-Controlled Physical Modelling Music Synthesis with Tactile Feedback', *Journal on Applied Signal Processing*, 7, 1001–6.

Hughes, D. (2017), 'Vocal Health Challenges and Contemporary Singing: Implications and Advocacy', *Australian Voice*, 18, 14–22.

Husserl, E. (1901), *Logical Investigations*, trans. J. N. Findlay, London: Routledge & Kegan Paul.

Husserl, E. (1962), *Ideas: A General Introduction to Pure Phenomenology*, trans. B. Gibson, New York: Colliers Macmillan.

Ihde, D. (2013), 'Embodiment: Technologies and Musics', in H. De Preester (ed.), *Moving Imagination: Explorations of Gesture and Inner Movement*, 101–12, Amsterdam: John Benjamins.

Ingebritsen, R., Knowlton, C., Sato, H., & Mott, E. (2020), 'Social Movements: A Case Study in Dramaturgically-Driven Sound Design for Contemporary Dance

Performance to Mediate Human-Human Interaction', in *Proceedings of the Fourteenth International Conference on Tangible, Embedded, and Embodied Interaction*, 227–37, New York: ACM.

Jensenius, A. R. (2007), 'Action-Sound: Developing Methods and Tools to Study Music-Related Body Movement', PhD diss., University of Oslo, Norway.

Jensenius, A. R. (2013), 'An Action–Sound Approach to Teaching Interactive Music', *Organised Sound*, 18(2), 178–89.

Jensenius, A. R., Wanderley, M. M., Godøy, R. I., & Leman, M. (2010), 'Musical Gestures Concepts and Methods in Research', in R. I. Godøy & M. Leman (eds), *Musical Gestures: Sound, Movement and Meaning*, 12–35, New York: Routledge.

Jessop Nattinger, E. (2014), 'The Body Parametric: Abstraction of Vocal and Physical Expression in Performance', PhD diss., Massachusetts Institute of Technology, Cambridge.

Johnson, M. (1987), *The Body in the Mind: The Bodily Basis of Meaning, Reason and Imagination*, Chicago: University of Chicago Press.

Johnson, M. (2007), *The Meaning of the Body*, Chicago: University of Chicago Press.

Johnston, A. (2009), 'Interfaces for Musical Expression Based on Simulated Physical Models', PhD diss., University of Technology, Sydney, Australia.

Jordà, S. (2003), 'Sonigraphical Instruments: From FMOL to the ReacTable', in *Proceedings of the International Conference on New Interfaces for Musical Expression*, 70–6, Montreal, Canada: McGill University.

Jordà, S. (2004), 'Instruments and Players: Some Thoughts on Digital Lutherie', *Journal of New Music Research*, 33(3), 321–41.

Kazlauskaite, S. (2020), 'Women Sonic Thinkers: The Histories of Seeing, Touching and Embodying Sound', in S. Krogh Groth & H. Schulze (eds), *The Bloomsbury Handbook of Sound Art*, 336–55, London: Bloomsbury Academic.

Kendon, A. (2004), *Gesture: Visible Action as Utterance*, Cambridge: Cambridge University Press.

Kjölberg, J. (2004), 'Designing Full Body Movement Interaction Using Modern Dance as a Starting Point', in *Proceedings of the Conference on Designing Interactive Systems: Processes, Practices, Methods, and Techniques 2004*, 353–6, Cambridge, MA: ACM.

Kozel, S. (2007), *Closer: Performance, Technologies, Phenomenology*, Cambridge, MA: MIT Press.

Kvifte, T. (2011), 'Musical Instrument User Interfaces: The Digital Background of the Analog Revolution', in *The International Conference on New Instruments for Musical Expression*, Keynote Talk, Oslo, Norway.

Kvifte, T., & Jensenius, A. R. (2006), 'Towards a Coherent Terminology and Model of Instrument Description and Design', in *Proceedings of the International Conference on New Interfaces for Musical Expression (NIME'06)*, 220–5, Paris: IRCAM.

Laban, R., & Lawrence, F. C. (1974), *Effort*, London: Macdonald & Evans.

Lakoff, G., & Johnson, M. (1980), *Metaphors We Live By*, Chicago: University of Chicago Press.

Lakoff, G., & Johnson, M. (1999), *Philosophy in the Flesh: The Embodied Mind and its Challenge to Western Thought*, New York: Basic Books.

Lane, C. (2020), 'Gender, Intimacy, and Voices in Sound Art. Encouragements, Self-portraits, and Shadow Walks', in S. Krogh Groth & H. Schulze (eds), *The Bloomsbury Handbook of Sound Art*, 198–212, London: Bloomsbury Academic.

Larssen, A. T., Robertson, T., Loke, L., & Edwards, J. (2007), 'Introduction to the Special Issue on Movement-Based Interaction', *Personal and Ubiquitous Computing*, 11(8), 607–8.

Lee, W., Lim, Y. K., & Shusterman, R. (2014), 'Practicing Somaesthetics: Exploring its Impact on Interactive Product Design Ideation', in *Proceedings of the 2014 Conference on Designing Interactive Systems*, 1055–64, New York: ACM.

Le Guin, U. K. ([1989] 2017), *Dancing at the Edge of the World: Thoughts on Words, Women, Places*, New York: Grove/Atlantic.

Lem, A., & Paine, G. (2011), 'Dynamic Sonification as a Free Music Improvisation Tool for Physically Disabled Adults', *Music & Medicine*, 3(3), 182–8.

Leman, M. (2010), 'Music, Gesture and the Formation of Embodied Meaning', in R. I. Godøy & M. Leman (eds), *Musical Gestures: Sound, Movement, and Meaning*, 126–53, New York, Routledge.

Leman, M. (2016), *The Expressive Moment: How Interaction (with Music) Shapes Human Empowerment*, Cambridge, MA: MIT press.

Leman, M., & Camurri, A. (2006), 'Understanding Musical Expressiveness Using Interactive Multimedia Platforms', *Musicae Scientiae*, 10(1), supplement, 209–33.

Leman, M., & Godøy, R. I. (2010), 'Why Study Musical Gestures?', in R. I. Godøy & M. Leman (eds), *Musical Gestures: Sound, Movement, and Meaning*, 3–11, New York, Routledge.

Levitin, D., McAdams, S. & Adams, R. L. (2002), 'Control Parameters for Musical Instruments: A Foundation for New Mappings of Gesture to Sound', *Organised Sound*, 7(1), 171–89.

Lewis, G. (2007), 'The Virtual Discourses of Pamela Z', *Journal of the Society of American Music*, 1(1), 57–77.

Linz, R. (2003), 'The Free Music Machines of Percy Grainger', *Rainer Linz*, available online: http://www.rainerlinz.net/NMA/articles/FreeMusic.html (accessed 3 October 2021).

Loke, L. (2009), 'Moving and Making Strange: A Design Methodology for Movement-Based Interactive Technologies', PhD diss., University of Technology, Sydney, Australia.

Loke, L., & Schiphorst, T. (2018), 'The Somatic Turn in Human-computer Interaction', *Interactions*, 25(5), 54–5863.

Lusted, H. S., & Knapp, R. B. (1998), 'Biomuse: Musical Performance Generated by Human Bioelectric Signals', *Journal of the Acoustical Society of America*, 84, S179.

Luo, X., & Hayes, L. (2019), 'Vibrotactile Stimulation Based on the Fundamental Frequency Can Improve Melodic Contour Identification of Normal-Hearing Listeners with a 4-Channel Cochlear Implant Simulation', *Frontiers in Neuroscience*, 13, 1145.

Machover, T. (2004), 'Shaping Minds Musically', *BT Technology Journal*, 22(4), 171–9.

Magnusson, T. (2019), *Sonic Writing: Technologies of Material, Symbolic, and Signal Inscriptions*, New York: Bloomsbury Academic.

Mailman, J. B., & Paraskeva, S. (2013), 'Continuous Movement, Fluid Music and Expressive Immersive Interactive Technology: The Sound and Touch of Ether's Flux', in M. Wyers & O. Glieca (eds), *Sound, Music and the Moving-Thinking Body*, 35–51, Newcastle upon Tyne: Cambridge Scholars,.

Mainsbridge, M. (2021), 'Soma-Based Non-Physical Instrument Design in Electronic Music Performance', *Leonardo*, 54(4), 393–7.

Malizia, A., & Bellucci, A. (2012), 'The Artificiality of Natural User Interfaces', *Communications of the ACM*, 55(3), 36–8.

Malloch, J. W. (2008), 'A Consort of Gestural Musical Controllers: Design, Construction, and Performance', PhD diss., McGill University, Montreal, Canada.

Mandanici, M., & Sapir, S. (2012), 'Disembodied Voices: A Kinect Virtual Choir Conductor', in *Proceedings of the Sound and Music Computing Conference*, 271–6, Copenhagen, Denmark: Aalborg University.

Márquez Segura, E., Turmo Vidal, L., Rostami, A., & Waern, A. (2016), 'Embodied Sketching', in *Proceedings of the 2016 CHI Conference on Human Factors in Computing Systems*, 6014–27, New York: ACM.

Maier, C. J., & Papalexandri-Alexandri, M. (2020), 'Membrane: Materialities and Intensities of Sound', in S. Krogh Groth & H. Schulze (eds), *The Bloomsbury Handbook of Sound Art*, 447–57, New York: Bloomsbury Academic.

Marshall, M. T., & Wanderley, M. M. (2011), 'Examining the Effects of Embedded Vibrotactile Feedback on the Feel of a Digital Musical Instrument', in *Proceedings of the 11th International Conference on New Interfaces for Musical Expression*, 399–404, Oslo, Norway: University of Oslo.

Mauss, M. (1973), 'Techniques of the Body', *Economy and Society*, 2(1), 70–88.

Max/MSP (n.d.), Computer Software. *Max*, https://cycling74.com/products/max.

McArthur, S. (2014), 'The Woman Composer, New Music and Neoliberalism', *Musicology Australia*, 36(1), 36–52.

McClary, S. (1998), 'Unruly Passions and Courtly Dances: Technologies of the Body in Baroque Music', in S. E. Melzer & K. Norberg (eds), *From the Royal to the Republican Body*, 85–112, Berkeley: University of California Press.

McNeill, D. (2005), *Gesture and Thought*, Chicago: University of Chicago Press.

Merleau-Ponty, M. (1964), *Signs*, trans. R. C. McCleary, Evanston, IL: Northwestern University Press.

Merleau-Ponty, M. (1968), *The Visible and the Invisible: Followed by Working Notes*, trans. A. Lingis, IL: Northwestern University Press.

Merleau-Ponty, M. (1999), *Phenomenology of Perception*, trans. C. Smith, London: Routledge.

Mewburn, I. B. 2009, 'Constructing Bodies: Gesture, Speech and Representation at Work in Architectural Design Studios', PhD diss., University of Melbourne, Australia.

Miranda, E. R., & Wanderley, M. M. (2006), *New Digital Musical Instruments: Control and Interaction beyond the Keyboard*, vol. 21, Middleton, WI: AR Editions.

Mitchell, T. J., Madgwick, S., & Heap, I. (2012), 'Musical Interaction with Hand Posture and Orientation: A Toolbox of Gestural Control Mechanisms', In *Proceedings of the 12th International Conference on New Interfaces for Musical Expression (NIME'12)*, Ann Arbor: University of Michigan.

Modalys (n.d.), Computer Software. 'Creating Virtual Instruments Based on Physical Models'. IRCAM, https://forum.ircam.fr/projects/detail/modalys/.

Moen, J. (2006), 'KinAesthetic Movement Interaction: Designing for the Pleasure of Motion', PhD diss., University of Stockholm, Sweden.

Moore, F. R. (1998), 'The Dysfunctions of MIDI', *Computer Music Journal*, 12, 19–28.

Montero, C. S., Alexander, J., Marshall, M. T., & Subramanian, S. (2010), 'Would You Do That? Understanding Social Acceptance of Gestural Interfaces', in *Proceedings of the 12th international Conference on Human Computer Interaction with Mobile Devices and Services*, 275–8, New York: ACM.

Morgan, F. (2014), 'Women's March', *The Wire*, 363, 53–4.

Morgan, F. (2017), 'Pioneer Spirits: New Media Representations of Women in Electronic Music History', *Organised Sound*, 22(2), 238–9.

Mulder, A. (2000), 'Towards a Choice of Gestural Constraints for Instrumental Performers', in M. M. Wanderley & M. Battier (eds), *Trends in Gestural Control of Music*, 315–35, Paris: IRCAM.

Murray-Browne, T., Mainstone, D., & Bryan-Kinns, N. (2011), 'The Medium Is the Message: Composing Instruments and Performing Mappings', in *Proceedings of the International Conference on New Interfaces for Musical Expression*, 56–9, Oslo, Norway: University of Oslo.

Murray-Browne, T., & Plumbley, M. (2014), 'Harmonic Motion: A Toolkit for Processing Gestural Data for Interactive Sound', in *Proceedings of the 14th International Conference on New Interfaces for Musical Expression (NIME'14)*, 213–6, United Kingdom: Goldsmiths, University of London.

Nafisi, J. (2015), 'Gesture and Body-Movement as Tools to Improve Vocal Tone', *Australian Voice*, 17, 11–21.

Naphtali, D. (2017), 'What If Your Instrument Is Invisible?', in *Musical Instruments in the 21st Century*, 397–412, Singapore: Springer.

Nelson, R. (2009), *The Jealousy of Ideas: Research Methods in the Creative Arts*, Goldsmiths, University of London.

Newlove, J., & Dalby, J. (2004), *Laban for All*, London: Routledge.

Nijs, L. (2017), 'The Merging of Musician and Musical Instrument: An Internal Model-Based Approach', in M. Lesaffre, P. -J. Maes & M. Leman (eds), *Embodied Musical Interaction*, 49–57, London: Routledge.

Noland, C. (2009), *Agency and Embodiment: Performing Gestures/Producing Culture*, Cambridge, MA: Harvard University Press.

Norman, D. A. (2010), 'Natural User Interfaces Are Not Natural', *Interactions*, 17(3), 6–10.

Norman, D. A., & Nielsen, J. (2011), 'Gestural Interfaces: A Step Backwards in Usability', available online: http://www.jnd.org/dn.mss/gestural_interfaces_a_step_backwards_in_usab ility_6.html (accessed 14 September 2021).

Nunez-Pacheco, C., & Loke, L. (2014), 'Crafting the Body-Tool: A Body-Centred Perspective on Wearable Technology', in *Proceedings of the 2014 Conference on Designing Interactive Systems*, 553–66, New York: ACM.

Oliveros, P. (2005), *Deep Listening: A Composer's Sound Practice*, New York: Universe.

O'Modhrain, S. (2000), 'Playing by Feel: Incorporating Haptic Feedback into Computer-Based Musical Instruments', PhD diss., Stanford University, California.

O'Modhrain, S. (2011), 'A Framework for the Evaluation of Digital Musical Instruments', *Computer Music Journal*, 35(1), 28–42.

O'Modhrain, S., & Gillespie, R. B. (2018), 'Once More, with Feeling: Revisiting the Role of Touch in Performer-Instrument Interaction', in S. Papetti & C. Saitis (eds), *Musical Haptics*, 11–27, Cham: Springer.

Overholt, D. (2009), 'The Musical Interface Technology Design Space', *Organised Sound*, 14(2), 217–26.

Paine, G. (2009), 'Towards Unified Design Guidelines for New Interfaces for Musical Expression', *Organised Sound*, 14(2), 142–55.

Paine, G. (2015), 'Interaction as Material: The Techno-Somatic Dimension', *Organised Sound*, 20(1), 82–9.

Pedrosa, R., & MacLean, K. (2008), 'Perceptually Informed Roles for Haptic Feedback in Expressive Music Controllers', in A. Pirhonen & S. Brewster (eds), *Haptic and Audio Interaction Design*, 21–9, Berlin: Springer.

Poikonen, H., Toiviainen, P., & Tervaniemi, M. (2018), 'Naturalistic Music and Dance: Cortical Phase Synchrony in Musicians and Dancers', *PloS One*, 13(4), e0196065. https://doi.org/10.1371/journal.pone.0196065.

Ponce, J. B. (2007), *Fractured Bodies: Gesture, Pleasure and Politics in Contemporary Computer Music Performance*, Ann Arbor, MI: ProQuest.

Puckette, M. (n.d.), 'Pure Data (Pd)'. Computer software. *Pure Data*, https://puredata.info.

Rico, J., Crossan, A., & Brewster, S. (2011), 'Gesture-Based Interfaces: Practical Applications of Gestures in Real World Mobile Settings', in D. England (ed.), *Whole Body Interaction*, 173–86, London, Springer.

Roddy, S., & Furlong, D. (2013), 'Rethinking the Transmission Medium in Live Computer Music Performance', *Irish Sound Science and Technology Convocation*, Dún Laoghaire Institute of Art, Design and Technology Laoghaire, available online: <http://issta.ie/wp- content/uploads/ISSTC-2013-RODDY.pdf>.

Rogers, T. (2010), *Pink Noises: Women on Electronic Music and Sound*, Durham, NC: Duke University Press.

Rokeby, D. (1998), 'The Construction of Experience: Interface as Content', in C. Dodsworth (ed.), *DigitalIillusion: Entertaining the Future with High Technology*, 27–47, Menlo Park, CA: Addison-Wesley.

Rokeby, D. (2010), *David Rokeby*, available online: http://www.davidrokeby.com/vns.html (accessed 22 March 2021).

Rominger, C., Fink, A., Weber, B., Papousek, I., & Schwerdtfeger, A. R. (2020), 'Everyday Bodily Movement Is Associated with Creativity Independently from Active Positive Affect: A Bayesian Mediation Analysis Approach', *Scientific Reports*, 10(1), 1–9.

Rovan, J., & Hayward, V. (2000), 'Typology of Tactile Sounds and Their Synthesis in Gesture-Driven Computer Music Performance', in M. M. Wanderley & M. Battier (eds), *Trends in Gestural Control of Music*, 297–320, Paris: IRCAM.

Saffer, D. (2008), *Designing Gestural Interfaces: Touchscreens and Interactive Devices*, Sebastopol, CA: O'Reilly Media.

Salter, C. (2012), '*JND*: An Artistic Experiment in Bodily Experience as Research', in D. Peters, G Eckel & A. Dorschel (eds), *Bodily Expression in Electronic Music: Perspectives on Reclaimed Performativity*, 181–99, New York: Routledge.

Schacher, J. C. (2012), 'The Body in Electronic Music Performance', in *Proceedings of the International Sound and Music Computing Conference*, 194–200, Copenhagen, Denmark: Aalborg University,.

Schiphorst, T. (2009), 'Body Matters: The Palpability of Invisible Computing', *Leonardo*, 42(3), 225–30.

Schiphorst, T. (2011), 'Self-Evidence: Applying Somatic Connoisseurship to Experience Design', in *CHI'11 Extended Abstracts on Human Factors in Computing Systems*, 145–60, Vancouver, Canada: ACM.

Schloss, W. A. (2003), 'Using Contemporary Technology in Live Performance: The Dilemma of the Performer', *Journal of New Music Research*, 32(3), 239–42.

Shackel, B. (1986), 'IBM Makes Usability as Important as Functionality', *Computer Journal*, 29(5), 475–6.

Sheets-Johnstone, M. (1999), *The Primacy of Movement*, Amsterdam, John Benjamins.

Sheets-Johnstone, M. (2010), 'Body and Movement: Basic Dynamic Principles', in D. Schmicking & S. Gallagher (eds), *Handbook of Phenomenology and Cognitive Science*, 217–34, Netherlands, Springer.

Sheets-Johnstone, M. (2013), 'Bodily Resonance', in H. De Preester (ed.), *Moving Imagination: Explorations of Gesture and Inner Movement*, 19–36, Amsterdam: John Benjamins.

Shusterman, R. (2009), 'Body Consciousness and Performance: Somaesthetics East and West', *Journal of Aesthetics and Art Criticism*, 67(2), 133–45.

Shusterman, R. (2012), *Thinking through the Body: Essays in Somaesthetics*, Cambridge: Cambridge University Press.

Shusterman, R. (2013), Affective Cognition: From Pragmatism to Somaesthetics', *Intellectica*, 60(2), 49–68.

Shusterman, R. (2020), 'Somaesthetics in Context', *Kinesiology Review*, 9(3), 245–53.

Sloboda, J. A. (2005), *Exploring the Musical Mind: Cognition, Emotion, Ability, Function*, Oxford: Oxford University Press.

Sonami, L. (2014), 'And Now We Leave Gloves and Other Wearables to (Small) Dictators', available online: https://sonami.net/?page_id=674 (accessed 22 March 2021).

Sonami, L. (2017), 'Requiem for the Lady's Glove', *Sound American: The Maker Issue*, 2(6), 139–43.

Spool, J. (2005), 'What Makes a Design Seem Intuitive?', *User Interface Engineering*, available online: http://www.uie.com/articles/design_intuitive/ (accessed 14 September 2021).

Spry, T. (2016), *Body, Paper, Stage: Writing and Performing Autoethnography*, New York: Routledge.

Spry, T. (2017), 'Who Are "We" in Performative Autoethnography?', *International Review of Qualitative Research*, 10(1), 46–53.

Svanæs, D., & Barkhuus, L. (2020), 'The Designer's Body as Resource in Design: Exploring Combinations of Point-of-View and Tense', in *Proceedings of the 2020 CHI Conference on Human Factors in Computing Systems*, 1–13, Honolulu, HI: ACM.

Tanaka, A. (2000), 'Musical Performance Practice on Sensor-Based Instruments', in M. M. Wanderley & M. Battier (eds), *Trends in Gestural Control of Music*, 389–405, Paris: IRCAM.

Tanaka, A. (2012), 'BioMuse to Bondage: Corporeal Interaction in Performance and Exhibition', in M. Chatzichristodoulou & R. Zerihan (eds), *Intimacy across Visceral and Digital Performance*, 159–69, Basingstoke: Palgrave Macmillan.

Tanaka, A. (2015), 'Intention, Effort, and Restraint: The EMG in Musical Performance', *Leonardo*, 48(3), 298–9.

Tanaka, A., Di Donato, B., Zbyszynski, M., & Roks, G. (2019), 'Designing Gestures for Continuous Sonic Interaction', in *Proceedings of the International Conference of New Musical Interfaces for Musical Expression*, 180–5, Porto Alegre, Brazil: Federal University of Rio Grande do Sul.

Tanaka, A., & Donnarumma, M. (2019), 'The Body as Musical Instrument', in Y. Kim & S. L. Gilman (eds), *The Oxford Handbook of Music and the Body*, 79–96, Oxford: Oxford University Press.

Tarvainen, A. (2019), 'Music, Sound, and Voice in Somaesthetics: Overview of the Literature', *Journal of Somaesthetics*, 5(2), 8–23.

Tarvainen, A., & Järviö, P. (2019), 'Preface: Somaesthetics and Sound', *Journal of Somaesthetics*, 5(2), 4–7.

Tassabehji, R., Harding, N., Lee, H., & Dominguez-Pery, C. (2020), 'From Female Computers to Male computǒrs: Or Why There Are so Few Women Writing Algorithms and Developing Software?', *Human Relations*, 0018726720914723.

Thompson, M. (2016), 'Feminised Noise and the "Dotted Line" of Sonic Experimentalism', *Contemporary Music Review*, 35(1), 85–101, DOI: 10.1080/07494467.2016.1176773.

Torre, G., & Andersen, K. (2017), 'Instrumentality, Time and Perseverance', in H. Egermann, S. I. Hardjowirogo, S. Weinzierl, T. Bovermann & A. de Campo (eds), *Musical Instruments in the 21st Century*, 127–36. Singapore: Springer.

Varela, F. J., Thompson, E., & Rosch, E. (1992), *The Embodied Mind: Cognitive Science and Human Experience*, Cambridge, MA: MIT Press.

Vines, B., Wanderley, M., Krumhansl, C., Nuzzo, R., & Levitin, D. (2004), 'Performance Gestures of Musicians: What Sructural and Emotional Information Do They Convey?', in *Gesture-Based Communication in Human-Computer Interaction: 5th International Gesture Workshop*, GW 2003, vol. 2915, 468–78, Berlin: Springer Verlag.

Visi, F., & Tanaka, A. (2021), 'Interactive Machine Learning of Musical Gesture', in E. R. Miranda (ed.), *Handbook of Artificial Intelligence for Music: Foundations, Advanced Approaches, and Developments for Creativity*, 771–98, Cham: Springer Nature.

Waisvisz, M. (2000), 'Round Table: Electronic Controllers in Music Performance and Composition', in M. M. Wanderley & M. Battier (eds), *Trends in Gestural Control of Music*, 616–55, Paris: IRCAM.

Wanderley, M. M. (2001), 'Performer-instrument Interaction: Application to Gestural Control of Sound Synthesis', PhD diss., Université Paris, France.

Wanderley, M. M. (2002), 'Quantitative Analysis of Non-Obvious Performer Gestures', in I. Wachsmuth & T. Sowa (eds), *Gesture and Sign Language in Human-Computer Interaction*, 241–53, Berlin: Springer-Verlag.

Wanderley, M. M., & Orio, N. (2002), 'Evaluation of Input Devices for Musical Expression: Borrowing Tools from HCI', *Computer Music Journal*, 26(3), 62–76.

Ward, N., Ortiz, M., Bernardo, F., & Tanaka, A. (2016), 'Designing and Measuring Gesture Using Laban Movement Analysis and Electromyogram', in *Proceedings of the 2016 ACM International Joint Conference on Pervasive and Ubiquitous Computing: Adjunct*, 995–1000, Heidelberg: ACM.

Warren, K. (2016), 'Composing and Performing Digital Voice Using Microphone-Centric Gesture and Control Data', in *Proceedings of the International Computer Music Conference (ICMC)*, 438–41, Utrecht, The Netherlands: Michigan Publishing.

Warren, K. (2018), 'Sound Technologies as Agency-Granting Prosthesis to Vocal Body', *Leonardo Music Journal*, 28, 30–3.

Wechsler, R. (2006), 'Artistic Considerations in the Use of Motion Tracking: Practices of Virtual Embodiment and Interactivity', in S. Broadhurst & J. Machon (eds), *Performance and Technology: Practices of Virtual Embodiment and Interactivity*, 60–77, Hampshire, Palgrave Macmillan.

Wessel, D., & Wright, M. (2002), 'Problems and Prospects for Intimate Musical Control of Computers', *Computer Music Journal*, 26(3), 11–22.

Westerkamp, H. (1990), 'The Breathing Room', *Électro Clips*, CD, Québec: Empreintes DIGITALes.

Wexelblat, A. (1995), 'An Approach to Natural Gesture in Virtual Environments', *ACM Transactions on Computer-Human Interaction (TOCHI)*, 2(3), 179–200.

Whitehouse. M. (1995), 'The Tao of the Body', in D. Johnson (ed.), *Bone, Breath and Gesture: Practices of Embodiment*, 241–52, Berkeley, CA, North Atlantic Books.

Whitehouse, M. S., & Pallaro, P. (2007), *Authentic Movement: Moving the Body, Moving the Self, Being Moved: A Collection of Essays, Volume Two*, London: Jessica Kingsley.

Whitten, S. (2017), The Opera Stage's Guest Blog on January 15, 2017, https://harvardchoruses.fas.harvard.edu/news/importance-core-training-singers.

Wilde, D. (2011), 'Swing That Thing: Moving to Move: The Poetics of Embodied Engagement', PhD diss., Monash University, Melbourne, Australia.

Wilkie, K., Holland, S., & Mulholland, P. 2010, 'What Can the Language of Musicians Tell Us about Music Interaction Design?', *Computer Music Journal*, 34(4), 34–48.

Wilson-Bokowiec, J., & Bokowiec, M. A. (2006), 'Kinaesonics: The Intertwining Relationship of Body and Sound', *Contemporary Music Review*, 25(1–2), 47–57.

Wishart, T. (1996), *On Sonic Art*, London: Routledge.

Wöllner, C., & Hohagen, J. (2017), 'Gestural Qualities in Music and Outward Bodily Responses', in C. Wöllner (ed.), *Body, Sound and Space in Music and Beyond: Multimodal Explorations*, 69–88, New York: Routledge.

Young, I. M. (1980), 'Throwing Like a Girl: A Phenomenology of Feminine Body Comportment Motility and Spatiality', *Human Studies*, 3(1), 137–56.

Z, Pamela (2000), 'A Tool Is a Tool', *Theater*, 30(2), 62–4.

Zhao, Jin-Lei, Jiang, W.-T., Wang, X., Cai, Z.-D., Liu, Z.-H., & Liu, G.-R. (2020), 'Exercise, Brain Plasticity, and Depression', *CNS Neuroscience & Therapeutics*, 26(9), 885–95. doi:10.1111/cns.13385.

Author Index

Subject Index

Note: Figures are indicated by page number followed by "f".

Printed in the USA
CPSIA information can be obtained
at www.ICGtesting.com
LVHW010320190823
755641LV00002B/201